First Steps in Statistics

First Steps in Statistics

Daniel B. Wright

Sage Publications
London • Thousand Oaks • New Delhi

First published 2002. Reprinted 2002

SAGE Publications Ltd
6 Bonhill Street
London EC2A 4PU

SAGE Publications Inc.
2455 Teller Road
Thousand Oaks, California 91320

SAGE Publications India Pvt Ltd
32, M-Block Market
Greater Kailash - I
New Delhi 110 048

British Library Cataloguing in Publication data

A catalogue record for this book is available from the British Library

ISBN 0 7619 5162 8
ISBN 0 7619 5163 6 (pbk)

Library of Congress Control Number 2001 132951

Typeset by SIVA Math Setters, Chennai, India
Printed in Great Britain by The Cromwell Press Ltd,
Trowbridge, Wiltshire

Contents

Illustrations

Figures

Tables

Acknowledgements

Thanks to everyone at Sage for their patience, in particular Ziyad Marar, Simon Ross and Michael Carmichael. Thanks also to several people for commenting on chapters, especially Soni Wright, and to the people who I talk statistics with, in particular David Routh and Andy Field. Thanks also to the several people who got emails from a strange person asking about their data and saying that he was thinking of using their study in a statistics book. With the exception of my own studies, which I reported because I know them really well, I picked studies that I felt would be of interest to students and also examples of great research. Finally, thanks to my students, who help me to teach.

1   Introducing Statistics

We use statistics throughout our daily lives. Usually these are not the 'formal' statistics that appear in statistics books and in scientific reports, but they are embedded, very innocently, in our conversations. Examples include saying that 'England will *probably* win the next World Cup' and 'it takes *about* twenty minutes to cook rice'. The aims of this book are to enhance your awareness of these natural language statistics, to allow you to translate these into 'formal' statistics and, in so doing, to enable you to conduct and interpret these statistics.

Consider the two examples mentioned above. Regardless of how good you think England's chances are for winning the next World Cup, you know roughly what the above statement means. When we use words like 'probably' we are not usually worried about the precise meaning of the phrase. Translating from natural language to formal statistics often involves becoming more precise. Here we might say that the *probability* of England winning is greater than 0.50 or 50%. Probability is at the heart of statistics and will be described throughout this book. If you had a standard deck of 52 cards, shuffled them thoroughly and were about to draw one card, the probability of it being red is 0.50. So using this analogy, the above statement means that it is more likely that England wins the next World Cup, than randomly choosing a red card from a well-shuffled deck of cards.

The second statement, 'it takes *about* twenty minutes to cook rice', is a statistical phrase because of the word 'about'. Depending on the amount and type of rice, the initial heat of the water, the type of stove and even the altitude at which you are cooking, the amount of time it takes to cook rice is not constant, but varies. Translating this into statistics it becomes 'twenty minutes is the *central tendency* for the time to cook rice, but the exact time may vary from this'. 'Central tendency' is what the statisticians would call what is written on the side of the rice box suggesting how long to cook the rice. It is the value that, across all situations, the rice manufacturers think is the best guess for proper cooking time. There are different and more precise ways of calculating the central tendency which will be considered later in this chapter.

For most of you, the main concern with regards to statistics is not to help you to become a better rice chef, but how statistics are used in the social and behavioural sciences. The point of these examples is to show how frequently statistics are encountered in our lives. During the course of your studies you will come across other 'everyday statistics' and also more formal statistics. This book describes various procedures for creating these statistics. In this first chapter, I introduce some basic concepts that will be used throughout the book.

Preliminary Concepts: Cases, Samples, Variables, Data and Population

Suppose that you were interested in the height of children in a primary school classroom. You might measure the height of 10 children in this classroom. This means that there are 10 *cases*. These cases are sometimes referred to by other names, like subjects, respondents or participants. These 10 cases are your *sample*. You will have measurements for 10 children's heights (Table 1.1).

The *variable* HEIGHT stands for the scores of these children. In the same way as the amount of time to cook rice varies across situations, a child's height depends on which child you are measuring. It is a convention in statistics to denote a variable with a name, and usually a name that gives some clue about the meaning of the variable. Here, HEIGHT is a sensible choice. We can denote the height for any case by putting the case number as a subscript after the variable name. Thus, $HEIGHT_4$ is the height measurement for the fourth case. In this example it is 97 cm. When referring to the variable, you should write $HEIGHT_i$. The subscript i tells the reader that the variable can take on different values for the different cases. For most studies you will have several variables. Here you might have $WEIGHT_i$, AGE_i and some nutritional information. All these variables

Table 1.1 **The heights of 10 children.**

	Case numbers									
	1	2	3	4	5	6	7	8	9	10
Height in cm	112	120	104	97	85	103	93	112	93	100

together are called the *data*. 'Data' is a plural noun; it refers to all the numbers ('datum', the singular, is seldom used).

In this example, we probably would not be *just* interested in the height of this sample of 10 children. We might be using the data from these children to *infer* something about all the children in the classroom. All the children in the classroom would be the *population*. It is even possible that we would be using information about these 10 children to infer something about *all* children of a certain age. This step, from data about a sample to inferences about a population, is the fundamental aspect of statistics. This will be discussed more thoroughly from Chapter 4 onwards.

Four Statistics

The remaining focus of this chapter is on four statistics: the mean, the median, the mode and the proportion. Each is a measure of central tendency and each describes some characteristic of the sample. They are used in different circumstances. At the end of this chapter I will describe some of the considerations that you should think about when deciding which of these statistics to use.

The Mean

Some words that are used in statistics have different definitions in English. The definition of 'mean' is not that of a particularly ill-tempered statistic. Instead it more closely refers to what, in English, we call the average. Consider the example of children's height. Three steps are required to calculate the mean of a set of numbers.

1 First, you have to count the total number of cases. The total number of cases in a sample is denoted with the letter n (sometimes capital N is used). Here $n = 10$ children.

2 Then you have to add up all the numbers. Here $112 + 120 + 104 + 97 + 85 + 103 + 93 + 112 + 93 + 100 = 1019$ cm.

If we had to write out all these numbers each time we wanted to say 'add up all of them' statistics books would be very long winded. Instead a summation sign, Σ, is used to stand for 'adding up all the values'. This is the capital Greek letter sigma. Greek letters are often used in statistics, although I will try to keep them to a minimum in this book. Sometimes you will see this written as

$$\sum_{i=1}^{n} \text{HEIGHT}_i = 1019 \, \text{cm}$$

which should be read as 'sum the values of HEIGHT_i from $i = 1$ to $i = n$' or simply 'the sum from HEIGHT_1 to HEIGHT_n'. Throughout this book, and most textbooks, the '$i = 1$' and 'n' parts are assumed and not written down. People usually just write ΣHEIGHT_i to mean sum all the values of HEIGHT_i.

3 Finally you divide this sum by the number of cases. The way that a mean is indicated is by drawing a line over the variable. If the variable was X_i, its mean is $\bar{X}$. For this example

$$\overline{\text{HEIGHT}} = \frac{\Sigma\text{HEIGHT}_i}{n} = \frac{1019 \, \text{cm}}{10 \, \text{children}} = 101.9 \, \text{cm per child}$$

So, we can say the mean height of these children is 101.9 cm per child.

A lot of the time you hear people talk about the average. This is what they are referring to. Averages are calculated for everything from batting averages to the average rainfall. Consider the following data for a week's rainfall in centimetres:

Sun	Mon	Tue	Wed	Thur	Fri	Sat
0	0	1	6	0	0	0

There are seven cases ($n = 7$ days). The sum of all the week's rainfall is 7 centimetres ($\Sigma\text{RAIN}_i = 7$ cm). The mean is 7 cm/7 days or 1.0 cm per day. This value would be important if you were interested in the level of water reservoirs, but it is not that informative if you were interested in booking a holiday and wanted to know what clothes to take. The following set, which gives the amount of rainfall for seven days in another location, produces the same mean:

Sun	Mon	Tue	Wed	Thur	Fri	Sat
1	1	1	1	1	1	1

So while you might include your sun glasses and sun block if packing for the first location, you might not for the second.

The Median

A few very high or very low points can have a large impact on the mean. In some cases this is good. Consider the rainfall example. If you were interested in reservoir levels, you would want Wednesday's 6 cm of rainfall to have a large impact. However, if you were a tourist you probably would want to know what the typical day was going to be like. The mean does not tell you this. The *median* is often used in these circumstances. It is the 'middle' point. There are the same number of points that are higher than it, as are points that are lower than it. The median is probably easiest to understand through an example. To find this value, arrange the points in order from the smallest value to the largest, and then locate the middle point. For the two sets of rain data, this would be the fourth point.

				Middle ↓			
Set 1	0	0	0	0	0	1	6
Set 2	1	1	1	1	1	1	1

When you have an odd number of cases, such as seven, the data point you look for is the $(n + 1)/2$ point. So with $n = 7$, it is $(7 + 1)/2 = 4$, or the fourth. Here the median for the first set is 0 cm, while the median for the second set is 1 cm. If you had 19 points, it would be $(19 + 1)/2 = 10$, the 10th.

When there is an even number of data points it is slightly more difficult. Consider the children's height data:

				Middle ↓				
85	93	93	97	100	103	104	112	112	120

With 10 data points there is no middle number. It is somewhere between the fifth and sixth point, between the $n/2$ point (here, $10/2 = 5$) and the $(n + 1)/2$ point (here, $12/2 = 6$). In these situations you should report the mean of these two numbers. This is called the *mid-rank* and is very important for the procedures discussed in Chapter 9. The mean of 100 and 103 is 101.5 cm. Therefore, this is the median height of these 10 children. This tells us that half of the children in the sample are shorter than 101.5 cm and half are taller.

The median is not highly influenced by extreme values, while the mean is. Suppose we had asked five people what their annual income was and got the following values:

£6k   £8k   £11k   £14k   £61k

Here the median is £11k. The mean, however, is £20k. If we were interested in what the typical person was making, the median is more appropriate. Suppose the

Table 1.2 **Frequencies and proportions for 'what comes to mind when science is mentioned?' (data calculated from Gaskell et al., 1993).**

	Physical science	Life science	Technology	Response Environmental science	Social science	Other	Total
Frequency	774	625	302	222	80	96	2099
Proportion	0.37	0.30	0.14	0.11	0.04	0.05	1.00

Note: The proportions are rounded to the hundredth place.

person making £61k was the chief executive of a recently privatised utility and by a stroke of luck had a pay rise to £161k. The mean goes up to £40k, even though four of the five people are unaffected. The median stays at £11k. If politicians talk about the average household income increasing, it is important to realise that they are probably using the mean and this is strongly influenced by those with the highest salaries.

The Mode and Proportions

The *mode* is the value that occurs most often. How often it occurs is its *proportion*. Gaskell, et al. (1993) asked 2099 people in the United Kingdom 'what comes to mind when science is mentioned?'. The responses are in Table 1.2.

The first row of numbers shows the *frequencies*. The frequency is the number of data points, or cases, with a particular value. Here 774 people said something to do with the physical sciences came to mind. This is the highest frequency and therefore 'physical science' is the mode. This is the most common response, but it does not mean that most of the people said it. It only means that more people said this than any of the other responses.

The *proportion* is the frequency of a response divided by the total sample size. For the physical sciences this is

$$\frac{774}{2099} = 0.37$$

The response, or value, with the highest frequency will also have the highest proportion. The proportion of 'physical science' responses is slightly larger than a third (0.33), but smaller than a half (0.50). It is 37% of the sample. When reporting the mode it is almost always advisable also to report the proportion for that modal value. The other proportions are calculated in the same way. For example, 80 of the 2099 people responded 'social science'. Its proportion is 80/2099 = 0.038 which rounds to 0.04 or 4%.

When to Use Each Measure

Several measures of central tendency were presented in the last section and other less common ones also exist. Each describes a different aspect of the data and each is more appropriate in different situations. There are two principal reasons for choosing one of these statistics over another. The first is the influence of extreme points. This is most applicable for deciding between the mean and the median. The mean is greatly influenced by extreme points. For example, if you are doing a survey of people's incomes a very high income will have a large impact on the mean, but a much smaller impact on the median. While sometimes researchers want these points to be highly influential, at other times they do not and would therefore use the median.

Mean   →   highly influenced by extreme points
Median   →   not highly influenced by extreme points

Extreme points are usually fairly rare. Therefore, the mean is more influenced by these rare occurrences than is the median. We say the median is *robust* because these rare occurrences do not alter its value as much. Several more robust statistics are discussed in Chapter 9.

The second reason relates to what is called the measurement level of the *scale*. This is a difficult concept (Wright, 1997a). When calculating the mean we are assuming a very particular relationship among the values. We assume what is called *interval scaling*. This is more easily explained through an example. For the height data (Table 1.1), we assumed that the difference between child 1 and 2 (8 cm) is the same as the difference between child 5 and 7 (8 cm). For interval data, it makes no difference where on the height scale that difference is. Intuitively this assumption seems fine for the height data. But suppose that it is not the height, *per se*, in which you are interested. Perhaps it is how tall somebody seems. I have not tested this, but I would guess that a difference of 8 cm seems much larger if the people were small children than if they were tall basketball players.

Consider the income example. If you were interested in income because you wanted to calculate the total income of a country, the difference between making £5k and £10k per year for two people is the same as the difference between making £105k and £110k per year. But most measures of wealth would consider the first difference to be larger. A shift from £5k to £10k will cause a much more noticeable shift in the standard of living than a shift from £105k to £110k. When you have a scale, like height or income, and you do not think that the interval assumptions are met, you have two choices. The first choice is to use the median rather than the mean. The second choice is to alter the data in some way so that they do seem to meet these assumptions. This is called *transforming* the data. A relatively simple transformation for the income example would be to treat the shift from £5k to £10k as the same as from £100k to £00k, a doubling of income. Transformations can become very complex and are, for the most part, beyond the scope of this book. However, one transformation is worth considering.

The simplest transformation is not to make any assumptions about the distances between values. Consider the income example presented earlier (£6k, £8k, £11k, £14k, £61k). This transformation involves *ranking* the data in order. List the data values in the order from smallest to largest:

Original data	£6k	£8k	£11k	£14k	£61k
Transformed (ranked)	1	2	3	4	5

It assumes what is called an *ordinal scale*. It was this level of measurement that was assumed when calculating the median, which also involved placing the data in order from smallest to largest.

Another measurement level is called *categorical* or *nominal*. Here the values cannot be ordered. The 'what comes to mind?' responses are an example of this. We would not assume that there is some scale or ordering in these values. They are just names (the word nominal comes from the Latin word *nomen*, for name). Other examples of categorical variables include 'the political party you plan to vote for', 'favourite instant coffee' and 'your religion'. There is a special type of categorical variable in which there are only two possible values: yes or no, male or female, true or false, etc. These are called *binary* variables. For categorical variables, the mode and proportions are appropriate.

Interval data	→	mean
Ordinal data	→	median
Categorical/binary data	→	mode/proportions

There are various issues that relate to these measurement levels and these issues are hotly debated in statistics. Here, I will simply note that the system is more flexible than I have described (see Box 1.1). In fact, the mode and proportions can be found for any type of data. For the income example, it would simply tell you which income had been earned by the most people. With most samples, this will be £0k. This does not mean that most people have no salary, only that it is more common than any other value. Similarly, the mode/proportions can be found for ordinal data. Finally, the median can also be used with interval data. If you are not sure which measure to report, you can report more than one. As each measure describes a different aspect of the data, and if you feel this aspect is important, each may provide additional and useful information. That is, providing that each makes sense. If you report, for example, that the mean response to the 'what comes to mind?' question is 'life science' you would be wrong and probably ridiculed by your classmates.

It is worth noting that when researchers are deciding between the mean and the median, they usually prefer working with means. Many statistical procedures have been developed to look at means. Because of this researchers usually look at the mean unless they have a reason not to. This said, there has been a welcomed increase in the past 10 years on using procedures that do not use the mean (Wilcox, 1998, discusses some recent advances. Chapter 9 of this book covers some of the more traditional alternatives).

**BOX 1.1   THE LEVEL OF MEASUREMENT DEPENDS
ON THE SITUATION**

Frederic Lord did some of the most important work on what is now called *item response theory*. In English, this is how to construct standardised tests and how to use people's responses to calculate scores. He also posed several interesting conceptual statistical questions and wrote about these in an entertaining way. Lord (1953) described two fetishes the mysterious Professor X had with numbers. First, he sold 'football numbers' to students at his university (the students may have been odd also). The football numbers on the backs of players' shirts are usually thought of as just nominal, not forming a scale. Professor X's second fetish was to collect ordinal-level data, but then calculate means on these data. While he performed his first fetish in public, much to the delight of his students, he hid in a darkened room, behind locked doors, to perform the unsightful (though strangely gratifying) task of calculating means on ordinal data.

Then his worlds collided. He sold his numbers to both first- and second-year students. Second-year students mocked their younger colleagues for having smaller numbers, and the freshers began a campaign of civil unrest. Professor X thought that the number sales were random. Could he be wrong, and how could he tell? He went to a statistician friend who promptly calculated the means by adding the numbers and dividing by the number of numbers. Professor X exclaimed 'You can't add them', to which the statistician retorted 'Oh, can't I? I just did.' Professor X, now furious at the statistician's behaviour (and in public, too), yelled 'Why, they aren't even ordinal scores.' To justify his apparently anti-social behaviour the statistician said 'The numbers don't know that [they are nominal] … since the numbers don't remember where they came from, they always behave just the same way, regardless.' After much introspection and playing with his numbers, Professor X unlocked his door and accepted the statistician's explanation.

The moral of this story is not that if you get too involved with statistics you develop strange fetishes (at least not automatically). The moral is the choice of using, for example, a mean versus a median, depends not just on the data, but on the research question posed. When the first- and second-year students at Professor X's university began quibbling about who had higher scores, a fairly legitimate way to answer this question is which group had the higher mean. This was despite 'football numbers' being a common textbook example for a categorical/nominal-level variable. If the question had been something else, the mean might not have been appropriate.

Summary

Statistics are used whenever we are uncertain about something. Much of our natural language can be translated into more formal statistical terminology. In a sense, the purpose of this book is to teach you a new language, *statisticalese*. Most of the complex equations that make up statistics textbooks can be translated into everyday English. *Statisticalese*, however, is more precise. Translating from English into statistical notation often requires you to be more precise about your own thoughts. There are several terms that were introduced in this chapter. It is worth learning this new vocabulary because these words will be used throughout the book, and used in lectures and journal articles.

Four statistics were introduced: the mean, the median, the mode and proportions. These will be discussed in the subsequent chapters. They are the basis of most of the statistics described in this book. In particular, statistical tests will be shown that allow you to compare these values for different groups and for different situations.

Exercises

1.1 What do the following symbols refer to: X_6, X_i, $\bar{X}$, Σ and n?

1.2 In Table 1.3 there is a (perhaps) hypothetical restaurant bill for 10 people. They decided to split the bill evenly for their dinners.

(a) How much should each person pay and what is the statistical name for this?

(b) If you were asked how much it costs for a typical meal at this restaurant, what amount would you give? What is the statistical term for it called?

(c) Steve ordered extra food to have at home. The cost of his meal was £6. He paid £7 for the extra food, so this should not be included in the amount

Table 1.3 **Meal prices at Dan's Curry Palace.**

Name	Price
Louise	£4.50
Steve	£13.00
Alice	£5.50
Simon	£4.50
Dave	£6.00
Joanne	£5.50
Mel	£5.00
Tom	£5.00
Andy	£5.50
Alexa	£6.00

paid by others. With this extra knowledge, what would be your answers for (a) and (b)?

1.3 Five friends meet at one of their houses to play poker. Each started with £20. They played for 10 hands (10 games). Each one of the players won at least one hand. No one else came in, and, except for the betting, no other money was spent. What was the mean number of pounds that each person had after these 10 hands?

1.4 The following exam marks, out of 100, were awarded to 20 students:

54 65 63 75 81 32 0 69 48 38 19 68 55 67 76 0 74 47 61 88

Find the mean, the median and mode of these data. For the mode, what proportion of the people received the modal grade? What do each of these measures tell you? Which do you think is best and why?

1.5 A drug company has created a drug that it feels can increase people's memory. The company had a group of 20 people whom it randomly allocated into two groups. One group of 10 people received a pill containing the drug while the other group received a pill without the drug (what is sometimes called a *placebo*). Participants were then given a memory test, scored from 0 to 10 where higher scores stand for having better memory scores. The data are:

Placebo group	2	5	4	4	1	3	4	10	4	4
Drug group	5	8	10	3	5	1	1	9	8	8

What are the means for the two groups? Does one group have a higher mean than the other?

1.6 My local sandwich shop has five choices of sandwiches: cheese, chicken, roast vegetables, ham and egg. I sat in the shop one lunchtime, doing what is called *participant observation*. This involved eating sandwiches for two hours straight (so that I would 'blend' in) and recording what other people were ordering. Not counting my own sandwiches, here are the sandwiches that people ordered:

13 people had cheese
23 people had chicken
8 people had roast vegetables
4 people had ham
14 people had egg

What statistic or statistics would you use to describe these data? What is/are the value(s) for it/them?

1.7 Find an 'everyday' statistical phrase from a newspaper. Say what the phrase means and why it is a statistical phrase.

Further Reading

Dorling, D., & Simpson, L. (Eds.) (1999). *Statistics in Society: The Arithmetic of Politics*. London: Arnold.
   This book contains chapters warning how statistics can be misused by governments and others. It warns that you should not always trust the statistics that you read.

Lord, F. M. (1953). On the statistical treatment of football numbers. *American Psychologist, 8*, 750–751.
   While this paper is described in Box 1.1, if you are in the library near the *American Psychologist*, please have a quick read of this paper. It is worth (a) seeing that I was telling the truth about the mad capers of Professor X, and (b) reading Lord's witty and enjoyable style. Sadly Lord past away in 2000.

Wright, D. B. (1997a). Football standings and measurement levels. *The Statistician: Journal of the Royal Statistical Society Series D, 46*, 105–110.
   Another paper on football and measurement, but this time European football (i.e. soccer). I talk about how the English FA Premiership method of tie breaking in the standings is based on the number of goals scored being interval data.

2 Graphing Variables

In the previous chapter some basic statistics were described. Many statistical techniques involve calculating a few numbers, like the mean or the median, to describe a variable. This approach often glosses over possible intricacies in the data. For example, in the previous chapter, rainfall in centimetres for a week from two hypothetical places was compared. These were

	Sun	Mon	Tue	Wed	Thur	Fri	Sat
Place 1	0	0	1	6	0	0	0
Place 2	1	1	1	1	1	1	1

The means for these were the same, 1 cm, but clearly the data say something very different about the rainfall in these two places. While the median, or the mode, can convey additional information, in some way these still will not show the full picture. The objective of this chapter is to teach you some techniques for displaying 'the full picture', for displaying data in graphs. Graphs often display the data much more clearly than a few statistical tests. Further, graphs often show aspects of the data which would not become apparent if only using statistical tests.

Throughout this book I will stress the importance of graphing data to complement statistical tests. Too often people think their statistics course involves just learning and using lots of equations. Graphs are a great way to communicate information about data.

Histograms and Distributions

One of the most useful statistical techniques is called a *histogram*. You often see these in newspapers, as well as in scientific reports. Let's consider an example. In many university departments, students get an overall mark for research methods, which includes both an essay-based exam on methodology and a statistics exam. As with most marking schemes, the different lecturers try to make the mean mark about the same for each part. Table 2.1 gives the data from a hypothetical course based roughly on the main UK university marking system where above 70 is a first-class degree (an 'A'), between 60 and 69 is an upper second (a 'B'), 50–59 a lower second (a 'C') and so on. In this example the mean for the essay exam is 58.53. The mean for the statistics exam is 57.51. Thus the means are about the same. But, reporting just the means hides a fundamental difference between essay and statistics exams. It is hard to do a great essay, but also hard to fail one badly. On the other hand, some people really have trouble with statistics and may do very poorly, while some do extremely well. In Table 2.1 each person has three data points. Using the first person as an example, these are 32 on the essay exam (the lowest mark; these are sorted by the essay marks, so the lowest mark is at the top left, the highest at the bottom right), and 14 on the statistics exam. The person is male ('M' for male, 'F' for female).

Figures 2.1, 2.2 and 2.3 are all examples of histograms using the data in Table 2.4. I will go through how to construct a histogram like the one in Figure 2.1 and then discuss how Figures 2.2 and 2.3 differ. It is worth pointing out that many computer packages can produce histograms, and if you have access to a computer it is worth learning how to make graphs on them. All the graphs in this chapter were made using SPSS for Windows, a popular statistical package. Most of the main word processing suites (e.g. Word, WordPerfect) allow graphs from other packages to be imported into documents and also often have graphing capabilities themselves.

Constructing a Histogram

1 Decide the Width of the Bars

To construct a histogram you first have to decide how many bars to have and where they should start. Because it is possible for the exam marks to range from 0 to 100, it seems that these points should be where the histogram begins and

Table 2.1 **Hypothetical essay (ESSAY) and statistics (STATS) exam marks. E stands for the essay mark, S for the statistics marks, and G for gender, M being male and F being female. These data are used in Figures 2.1–2.4, and later in this book.**

E	S	G	E	S	G	E	S	G	E	S	G	E	S	G
32	14	M	49	76	F	56	80	M	62	83	F	67	56	F
37	14	M	50	65	F	56	41	F	62	86	F	67	95	F
37	39	F	50	71	F	57	36	M	62	2	F	68	24	F
39	49	M	51	91	M	57	79	M	63	91	F	69	95	F
41	17	F	52	19	F	57	67	M	64	27	F	69	80	M
42	68	F	52	10	F	57	21	F	64	80	F	69	65	F
42	60	M	52	94	M	58	33	F	65	94	F	69	29	F
42	0	M	53	12	F	58	68	F	65	86	M	71	86	F
42	28	M	53	55	M	58	28	F	65	17	F	71	34	F
42	90	F	53	8	M	58	19	M	65	29	M	71	86	F
43	5	M	53	62	F	58	95	F	65	95	F	73	56	F
44	84	F	53	71	F	58	82	F	66	88	F	74	0	F
44	80	M	54	87	M	59	90	F	66	29	M	75	82	F
44	74	M	54	30	F	60	0	F	66	0	F	76	73	M
45	59	M	55	89	F	60	84	F	67	96	F	76	98	F
47	5	M	55	24	M	60	76	M	67	79	F	77	99	M
48	65	M	55	35	M	60	59	M	67	100	F	78	34	F
48	14	M	56	89	M	61	0	F	67	95	M	79	86	F
48	74	F	56	77	M	61	56	M	67	96	M	79	85	M
49	84	M	56	53	F	61	44	F	67	89	M	85	27	M

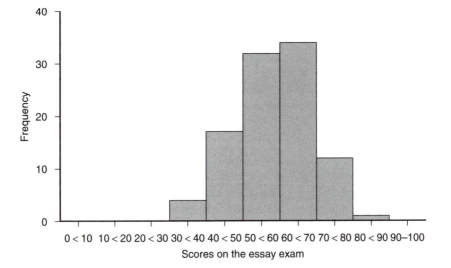

Figure 2.1 **A *histogram* showing the distribution of scores on the essay exam (ESSAY). The scores are divided up from 0 to 9, 10 to 19, etc., until 90 to 100. This shows that scores were centred around the mean of 58.53. See text for how to construct this graph.**

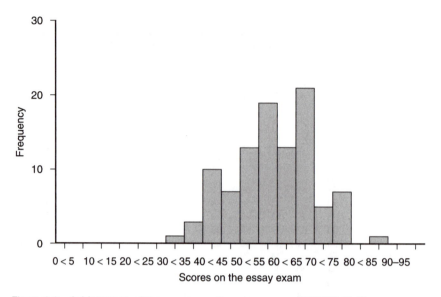

Figure 2.2   **A *histogram* of the scores on the essay exam (ESSAY). Unlike Figure 2.1, here scores are divided into five-point bands. These are 0 to 4, 5 to 9, until 95 to 100.**

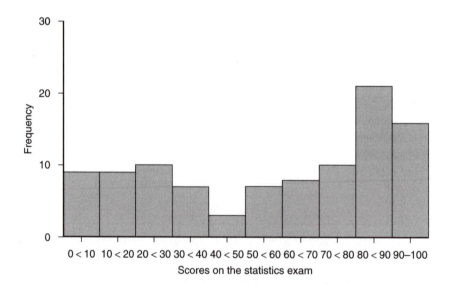

Figure 2.3   **A *histogram* of the scores on the statistics exam (STATS). The scores are in 10-point bands, like Figure 2.1. The mean is 57.51, which is similar to the mean for Figure 2.1, although the distributions are very different.**

ends. Further, because marks ('A', 'B', 'C', etc.) correspond to 10-point bands these seem the appropriate size to make the bands. So, for example, there will be a band from 60 to 70. Actually it goes *almost* to 70. If the grading was so precise that a student could get 69.9, that student would be in this band. However, a student getting 70.0 would not because that student would fall in the next grade up (an exception is 100 which is still in the 90–100 band). Often it is not so clear how many bands there should be. If you use a computer it will often choose for you, but the package should allow you to alter the number in order to suit your needs. In general it is worth thinking how to make your graphs as informative and useful for the reader as possible.

2 Calculate the Frequency for Each Bar

The next step is calculating the number of scores in each 10-point band. Table 2.1 is arranged so that the essay scores are in order. This makes the task easier. There were no scores below 30 so the bands 0 < 10, 10 < 20, and 20 < 30 have zero people in them. Four people scored in the 30s. We say the *frequency* for that band is 4. The 40s have 17 people, the 50s have 32, the 60s have 34, the 70s have 12 and there is 1 person in the 80s. No one scored in the 90s. These are the frequencies for each of the bands.

3 Draw and Label the Axes

Now, you actually draw the graph. This can be made easier by using graph paper and a ruler. First, draw a horizontal line like the line in Figure 2.1 that is above the names of the bands. When using graph paper this line should be along one of the lines already on the paper. Next draw a vertical line whose bottom meets the leftmost side of the horizontal line. Again, with graph paper this should be along one of the existing lines. These two lines are called the *axes* (axis being the singular). The horizontal one is called the *x axis* and the vertical one the *y axis*. Next, divide the *x* axis into 10 equal segments, or into however many bands that you are using. If you are using graph paper, you will probably want to have one or two of the boxes for each band. If you are not using graph paper then you will have to measure these carefully. Write the name of each band below the *x* axis. In Figure 2.1, I have used, for example, 60 < 70 to stand for 60 up to, but not including, 70. Next, put a title for the variable below these labels; make sure that the title is an appropriate description. 'Scores on the essay exam' seems a good description.

Now the *y* axis should be labelled. First, you have to find the band that has the highest frequency. There are 34 people who scored in the 60s. You need to have the top number on the *y* axis bigger than this. I chose 40 because it is a nice round number. The lowest point on the y axis will always be zero for a histogram. You then divide the y axis into convenient band breaks. I chose to break it into 10s. If you are using graph paper these should match up with the lines on the paper. For example, 40 could be 20 squares above the x axis. So 30 would make 15 squares above, 20 would make 10 above and 10 would be 5 squares above the x axis. Then label the axis with the word 'Frequency.'

4 Making the Bars

Now the main part of the graph can be made. Start at the leftmost band on the x axis that has a non-zero frequency. This is the $30 < 40$ band which has a frequency of four. Go to the y axis and find where four would occur. If you are using graph paper and have the number 40 marked 20 squares above the x axis, then each square stands for two people. Because this band has a frequency of four, it would be two squares above the x axis. With a ruler, go to this height directly above the $30 < 40$ band (on the x axis). Draw a horizontal line at this height that is the length of the band. Next draw vertical lines down from each end of this to the x axis. You will get a rectangle for the frequency of that band.

You then repeat this with each band. You may have to estimate the height. How much care and precision you use depends on what the graph is for. If you are just doing it for yourself and want to get a rough idea of what the data look like, then just try to get it *about* right. If the graph is for an important assignment then your precision should reflect this. Once all the rectangles have been drawn, you might want to fill in the rectangles to make them easier to see. Mine are filled in with grey. Finally, you should give a title and a number to the graph. I have done this in the caption below the graph. It is worth looking at Figure 2.1 if you have any trouble following these directions.

Histograms show the *distribution* of variables. Distribution essentially means the shape of the histogram, so the two are closely related. Figure 2.1 shows the distribution of the variable ESSAY using 10-point bands. Figure 2.2 shows what happens if you use five-point bands rather than 10-point ones. The conclusions from it are similar: almost all scores are between 40 and 80, and most are between 50 and 70.

The graph in Figure 2.3 shows the scores on the statistics exam, the variable STATS. It is very different. Here, using 10-point bands, it is clear the data are more spread out. Approximately two-thirds of the people scored less that 40 or greater than 80. In the next chapter there will be more discussion about describing the spread of a distribution. In addition to being spread out, the distribution is not 'bell shaped.' There appear to be two groups, one scoring low and one scoring high. This is called a *bi-modal* distribution.

'Bell-shaped' curves have a special place in statistics, so it is worth briefly mentioning them here. Figure 2.4 shows what is called the normal distribution

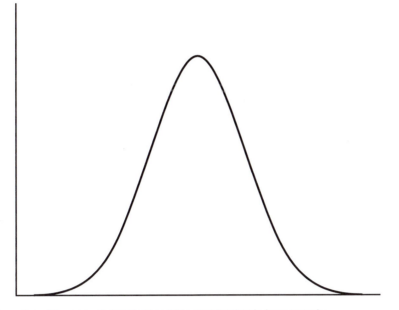

Figure 2.4   **The normal distribution. This distribution is important in statistics and will be discussed further in later chapters.**

(sometimes called the Gaussian distribution). It is shaped like one of those old-fashioned bells. In many of the statistical tests covered later in this book, it is assumed that they follow roughly this distribution. In reality most do not (Micceri, 1989).

Scatterplots

So far we have been looking at a single variable at a time, ESSAY *or* STATS. Suppose we were interested in how these two variables relate to each other. For example, we might be interested in whether people who do well in the essay exam also do well in the statistics exam. To explore this we make a *scatterplot* (sometimes called a *scattergram*). An example is shown in Figure 2.5, with the scores on ESSAY shown on the *x* axis and the scores on STATS shown on the *y* axis.

As with the histogram you begin by drawing the *x* and *y* axes, adding numbers to them and giving them titles. You have to decide which variable goes onto which axis. Here the choice is fairly arbitrary. The convention is: if you think one variable either causes another or is predictive of another, you put the one you are trying to predict on the *y* axis and the one you are using to predict it on the *x* axis. For example, if your two variables were the daily lunchtime temperatures and the number of ice creams bought in a local park, temperature would probably be on

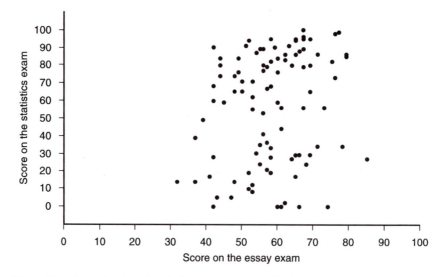

Figure 2.5 **A *scatterplot* of statistics exam score (STATS) with essay score (ESSAY).**

the *x* axis and ice creams on the *y* axis. This is because it is more likely that you would be using temperature to predict ice cream sales (if you were an ice cream vendor deciding how much stock to bring to the park) than vice versa. In Figure 2.5 I put the statistics scores on the *y* axis and the essay scores on the *x* axis, but this choice was arbitrary. You also have to decide where each axis begins and ends. Here I let the *x* axis (ESSAY) go from 0 to 100, since that is the possible range of scores for that variable. For the *y* axis, I let it run from just below 0 to just above 100. This is because there are scores of 0 and of 100 on this test. The corresponding points would be difficult to see otherwise. The axis numbers, their titles and the overall figure title and number are added in the same way as with Figure 2.1.

The final step is to add in the points, which can be done by making dots with a pen. The first person in Table 2.1 had 32 for ESSAY and 14 for STATS. You need to go along the *x* axis until you get to 32 then go up to 14 on the *y* axis. With 100 people in this sample this can become very time consuming. In Exercise 2.2 you have fewer points to include. In practice you would often have a computer do it for you. Most statistical packages also allow you to add extra information. For example, instead of using a dot, you could tell the computer to put an 'M' for each male and an 'F' for each female.

What can be told from this graph? Well, the data appear fairly spread out. There are a number of people who do fairly well on the essay but very poorly on the statistics. Conversely there are people who do very well on the statistics but are doing worse than most on the essay. However, there is a fairly large group of

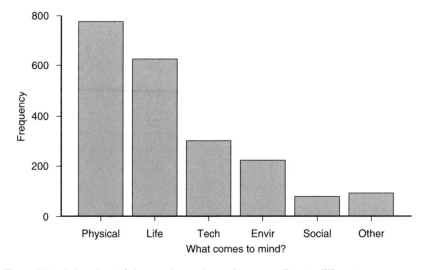

Figure 2.6 **A *barchart* of the numbers of people responding to different categories for 'what comes to mind' when science is mentioned. The data were calculated from Gaskell et al. (1993).**

people who do fairly well on both. There is a statistical technique, called *regression*, which helps to describe this relationship. It will be introduced in Chapter 8.

Barcharts

Histograms and scatterplots are both used when the variables are *quantitative*. You were introduced to the *interval* level of measurement in Chapter 1. This is one type of quantitative variable. A quantitative scale is simply one where it makes sense to talk about how far away from each other the values are. Also in the first chapter *categorical* (or nominal) variables were introduced. These are ones where there is no sense in comparing how far away different values are. An example from Gaskell et al. (1993) was used. They asked 2099 people 'what comes to mind?' when science is mentioned and found the following breakdown of responses:

Physical science	774
Life science	625
Technology	302
Environmental science	222
Social science	80
Other/don't know	96

These do not make a scale like exam scores. Sometimes it is best simply to report the numbers, as above, but at other times graphical techniques are useful in displaying this type of information. The most common technique in this situation is a *barchart*. One is depicted in Figure 2.6.

The steps for making a barchart are similar to those used when making a histogram. First you draw in the axes. Here you do not have to decide what scale to place on the *x* axis; all the possible values should be placed on the *x* axis. In the same way as with the histogram, you need to decide what the scale should be for the *y* axis, and label both axes. Since the highest frequency was 774 (for physical science), I chose 800 (see top of *y* axis). Now, as you did for each band in the histogram, make a rectangle showing the frequency for each option.

Summary

Graphs are becoming more common in statistics these days. Twenty-five years ago, most of the statistics described in this book could be calculated without too much trouble on fairly rudimentary computers. The computers in those days were 'mainframe' computers and most had poor graphic capabilities. Most statistical packages could produce only simple graphics. This led many social scientists into just doing the equation side of statistics and not exploring their data through graphs. Even these days, researchers often go straight into complex statistical techniques. It is as if something tempts researchers to conduct complex statistical procedures without actually looking at the data. Shame on them.

Advances in computing have meant that making graphs is much easier. Because personal computers became popular as entertainment systems rather than scientific aids, they have developed great graphics capabilities. Statistical packages started adding more and more sophisticated graphing procedures. You should *always* explore your data before running complex statistical analyses.

The three graphical techniques described in this chapter are common procedures for doing this. Each has variants, like using 'M' and 'F' to denote gender in Figure 2.5, that can make graphs even more useful. I will describe throughout this book other graphical techniques and more extensions of these three.

Graphs can be used both to explore data, called *exploratory data analysis* (EDA), and to display the information to others. This is often called *data display* or *data presentation*. There is one steadfast goal for both of these:

> Graphs are for presenting data as clearly and as accurately as possible. Their purpose is not to be artistic, though a good informative graph has its artistic appeal.

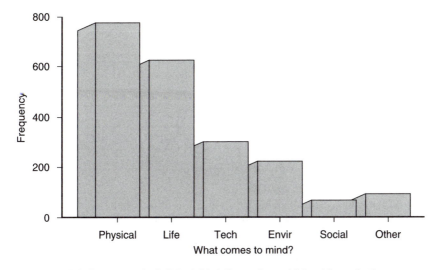

Figure 2.7 **A heinous graph. A 'false' third dimension, which adds no further information, has been added to the graph in Figure 2.6. This third dimension can easily confuse the reader.**

There are a few things that prevent this goal (see Wainer, 1984, for an excellent description of making graphs). One way to prevent reaching this goal is to add extra information that is not informative. The example of this that I most despise is adding a 'false' third dimension to a graph. Figure 2.7 presents the same data as Figure 2.6, but I have added worthless information. I made the bars three dimensional. This is at best a waste of ink and is likely to make the graph more difficult to read. Many graphics packages include this repulsive option. Wallgren and colleagues (1996: 71) suggest this is included because it 'symbolise[s] the triumph of data technology'. It is included simply to impress the user with the marvels of computer science rather than for any scientific purpose. When I politely criticise people for adding a false dimension to a graph, they often tell me it looks 'pretty'. 'Pretty' does not count. Information, clarity and accuracy count. 'Pretty' is fine providing these other three are fulfilled, but this and all false 3-D graphs are not even 'pretty'. They are repugnant. If I were not such a polite person I would ask, at the end of a talk by one of these 'pretty' people, what the third dimension represents. The person would have to admit that it meant nothing and that they were just trying to prove that they could press the 3-D button on the computer package.

You have probably noticed that I have a pet peeve, false 3-D graphs. There are lots of other ways that graphs can be made uninformative, unclear and inaccurate. Howard Wainer (1984) has written an entertaining and readable paper discussing these. In general, look at your graphs and ask yourself: 'is this the best way to communicate the information clearly and accurately?' Make sure that all the information you want to display is displayed, and that no part of the graph is uninformative.

Exercises

2.1 Following the instructions earlier this chapter, and using the data in Table 2.1, make a histogram of the variable ESSAY. It should look similar to the one in Figure 2.1. If you are using a computer, you may have to override some of the defaults.

2.2 When you see those advertisements from animal shelters and a dog with big, sad eyes seemingly says 'take me home', they really mean it. Hennessy and colleagues (1997) examined stress in dogs who recently arrived at a shelter by measuring their plasma cortisol levels. Suppose 10 dogs were sampled, their cortisol level measured in ng/ml was taken[1], and it was recorded how long they had been at the shelter. Pet dogs in people's homes have plasma cortisol levels of about 10 ng/ml. Based on Hennessy et al., below are data that might be obtained.

Plasma cortisol (ng)	24	32	15	18	16	24	34	22	28	19
Days in shelter		1	2	6	8	14	1	5	7	4	10

Make a scatterplot showing the relationship between days in the shelter and cortisol levels. Justify the decisions you had to make in constructing this scatterplot, and describe what the plot shows.

2.3 There is a great interest in child care. Melhuish (1991) was interested in differences in child care by mothers' social class in London. Consider the following data which he received from 250 women regarding where the child was being cared for.

at Home	70
with Relative	56
by Childminder	88
at Nursery	36

Make a barchart showing these data. Describe what this chart shows. What is the mode?

---

1 ng/ml is nanograms per millilitre. A nanogram is 0.000000001 of a gram and a millilitre is 0.001 of a litre. Basically, it is the concentration of cortisol in the blood.

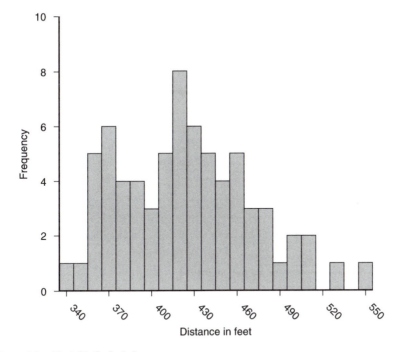

Figure 2.8 **Mark McGwire's home runs.**

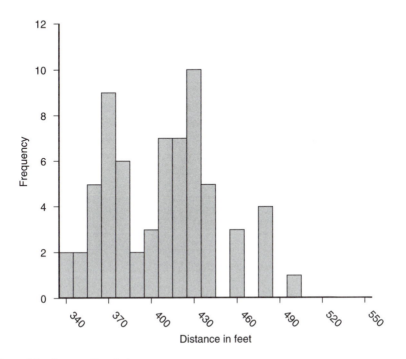

Figure 2.9 **Sammy Sosa's home runs.**

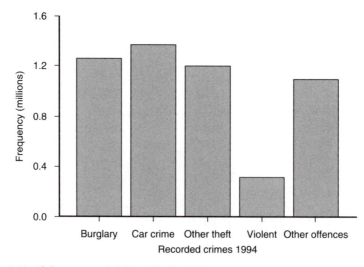

Figure 2.10   **Crimes recorded by police in 1994.**

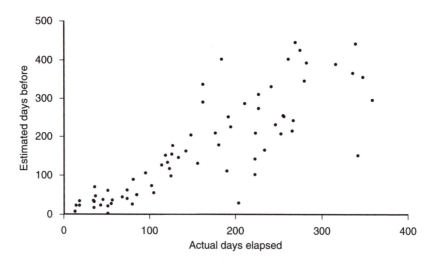

Figure 2.11   **A scatterplot showing estimated elapsed time (in days) since an event with the actual number of elapsed days.**

2.4   A 341 foot line drive down the left field line. Mark McGwire hits his 62nd home run of 1998, breaking Roger Maris's record and becoming part of sporting folklore. He says to the crowd: 'To all my family, my son, the Cubs, Sammy Sosa, it's unbelievable! Thank you, St. Louis!' (Schreiber, 1998: 83).

The home runs of Mark McGwire and Sammy Sosa did more than just bring life into a sporting public disenchanted with spiralling salaries and strikes, it provided two statisticians with data for showing how to make certain types of graphs. Keating and Scott (1999) demonstrate how to make some complex histograms that are beyond the scope of this text, but Figures 2.8 and 2.9 are the histograms for the distance that Mark McGwire's 70 and Sammy Sosa's 66 home runs travelled (McGwire's shortest was the record breaker). Describe these distributions. Who tended to have longer home runs?

2.5 Television crime programmes give the impression that violent crime is the norm. The barchart in Figure 2.10 shows the crime statistics for recorded crime in England and Wales in 1994. What does this tell you? It is probably worth mentioning that of the violent crimes, 64% were 'minor woundings'.

2.6 In surveys, people are often asked to say when some event occurred. For example, they might be asked when they last visited the dentist. The scatterplot in Figure 2.11 shows the actual days since an event (x axis) and the estimated number of elapsed days (y axis). Describe what this scatter-plot shows.

2.7 Graphs are very common in newspapers and magazines. Often the people creating these graphs try to make them 'pretty', rather than clear and accurate. Look through a newspaper or magazine and find a graph that you feel could be improved. Suggest how improvements could be made. Wainer's (1984) paper gives some good examples of bad graphs and how to improve them.

Further Reading

Everitt, B. S. (1999). *Making Sense of Statistics in Psychology: A Second-Level Course.* Oxford: Oxford University Press.
  This is a much more advanced textbook than mine. However, his chapter on graphs (Chapter 2: Graphical Methods of Displaying Data) is good and is probably appropriate 'next' reading after this chapter.

SPSS.COM.
  This is the website for two of the best general statistics of social sciences packages, SPSS and SYSTAT.

Wainer, H. (1984). How to display data badly. *American Statistician, 38,* 137–147.
  This article goes through several published figures and shows how they can be misleading, and also how to improve them. The style of the paper is great, and despite being in a statistics journal is readable.

Wainer, H., & Velleman, P. F. (2001). Statistical graphics: Mapping the pathways of science. *Annual Review of Psychology, 52,* 305–335.

    This paper describes the history of and best practices of making graphs.

Wilkinson, L. (2000). Cognitive science and graphic design. In *SYSTAT® 10 Graphics* (pp. 1–18). Chicago: SPSS Inc.

    Statistics manuals can be unfriendly dry reads. The SYSTAT ones are not. SYSTAT is great on graphics and Leland Wilkinson is great explaining the do's and don'ts.

3 The Spread of a Distribution

In the first chapter you were introduced to measures of central tendency. These describe the middle of a variable's distribution. In several examples it was clear that while two variables may have the same mean, median and/or mode, they may still differ in important ways. In the second chapter you were shown how to construct graphs. In this chapter I describe measures of the *spread* of a variable. Statisticians use the words 'spread' and 'dispersion' to describe how varied a variable's distribution is. Consider the difference between essay marks and statistics marks from Table 2.1 in Chapter 2. While the means were nearly identical, the marks for the statistics exam were more spread out than were the marks for the essay. Here I introduce ways to describe that spread. The four most used measures are the *range*, the *interquartile range*, the *variance* and the *standard deviation*.

It is worth noting that many people think statistics are just about the central tendency, for example if the means of two variables differ. The dispersion of a variable is often considered uninteresting. However, sometimes it is extremely informative. Further, the dispersion of a variable is used throughout the remainder of this book to give an indication of the precision of the estimates of the central tendency.

The Range and Interquartile Range

One way to describe dispersion is the *range* of a variable. The range is the distance between the highest and the lowest values. With the essay-based exam

(ESSAY) considered in Chapter 2, the highest score, or *maximum*, was 85. The lowest, or *minimum*, was 53. The range is $85 - 32 = 53$. For the statistics-based exam (STATS) the high score was 100 and the low score was 0, so the range is much larger for the statistics-based exam. The difference between the top and bottom score for the statistics exam is about twice the size of the difference for the essay exam.

The range has limitations because it can be greatly increased by a single high or low value. Ideally any statistic you report would not be greatly influenced by a single point. A simple alternative to the range statistic is called the *interquartile range* (sometimes called the mid-range and abbreviated IQR). The data for the essay exam are written below in the order from lowest to highest (values that are the same can be put in any order since they will be the same). Most computer packages will easily allow this. I have then divided the data into four equal sections. These sections are called *quartiles*. Because there are 100 students in this data set it is easy to do: the lowest 25 are in the first quartile, students 26 to 50 are in the second, 51 to 75 in the third, and the final 25 in the fourth quartile.

Essay scores (ESSAY)
First quartile   32, 37, 37, 39, 41, 42, 42, 42, 42, 42, 43, 44, 44, 44, 45, 47, 48, 48, 48, 49, 49, 50, 50, 51, 52
Second quartile   52, 52, 53, 53, 53, 53, 53, 54, 54, 55, 55, 55, 56, 56, 56, 56, 56, 57, 57, 57, 57, 58, 58, 58, 58
Third quartile   58, 58, 59, 60, 60, 60, 60, 61, 61, 61, 62, 62, 62, 63, 64, 64, 65, 65, 65, 65, 65, 66, 66, 66, 67
Fourth quartile   67, 67, 67, 67, 67, 67, 67, 68, 69, 69, 69, 69, 71, 71, 71, 73, 74, 75, 76, 76, 77, 78, 79, 79, 85

The IQR is the distance from the highest of the first quartile to the lowest of the fourth quartile. These two points are often called *hinges*. Here the interquartile range is $67 - 52 = 15$. This value is not as affected by a few extreme points as is the range. As such, statisticians say it is more *robust*.

Sometimes the number of cases cannot be divided into four equally sized groups. Further, there are some subtle differences in how some books define the IQR (e.g. with the above data, some books say that the IQR goes from the lowest of the second quartile to the highest of the third; this does not make a difference with this data set since these values are the same, but it would for the statistics scores). Many statistics computer programs, for example SPSS and SYSTAT, calculate the hinges and the IQR. These programs may use more complex equations but the concept is the same (and the different methods usually produce similar numbers). It is the concepts that are important; the IQR includes the middle 50% of the data. For consistency, here the IQR should be from the highest of the first quartile to the lowest of the fourth quartile.

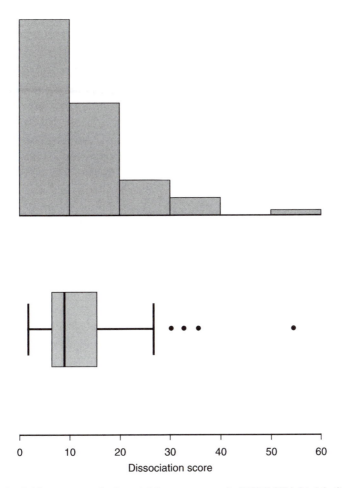

Figure 3.1 **A histogram and a boxplot for responses to DES II (Wright & Loftus, 1999), a questionnaire measuring dissociative experiences.**

The values for IQR, or more specifically the hinges, are used in a popular graphical technique called a *box and whiskers plot*, or sometimes just a *boxplot*.[1] There are some differences among textbooks on how to make these. Rather than get caught up in relatively minor differences, I will describe what the different parts of a boxplot are. Box 3.1 provides rules for making a relatively simple version of a boxplot. Boxplots show the median, the IQR and the values for *outliers*, and can be very helpful when comparing groups or variables. Figure 3.1 shows a

---

1 Technically there are some differences between these, but since they are often used interchangeably, I will keep with a basic boxplot.

BOX 3.1   MAKING A BOXPLOT BY HAND

There are some slight variations that can be used when constructing a box-plot. Here are the rules for a fairly simple version. Before beginning this you should have sorted all the data in order from lowest to highest, found the median, the upper and lower hinge, and the IQR.

1   On graph paper, make the horizontal axis so that the values cover the range of values in the same way as was done for histograms in Chapter 2.
2   Draw short lines, above the axis, to denote the medians for the different groups (or the median if you have a single group).
3   Draw a short vertical line for each hinge, and join these as a box (see Figure 3.1).
4   Calculate the values for the whiskers as follows:

Lower whisker = lower hinge − 1.5 × IQR
Higher whisker = higher hinge − 1.5 × IQR

Draw these as lines extending from the box to the further data point that is within this inner fence.
5   List any data points outside these whiskers with a star or some other character.

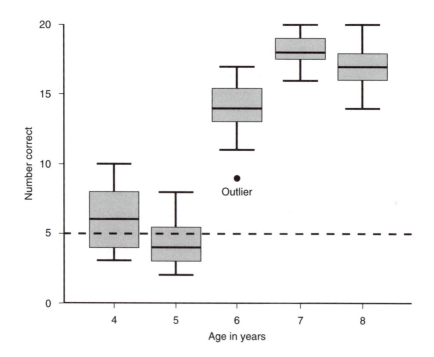

Figure 3.2   **Boxplots for the number of correct answers on a test by the participants' age. The dashed line shows the prediction for random guessing.**

histogram with the corresponding boxplot printed below it. The data are from 62 participants who were given a scale (called DES II) to measure the frequency of their dissociative experiences (Wright & Loftus, 1999). Dissociation refers to difficulty in integrating thoughts, memories and imagination, so people who dissociate may confuse imagined events with real events, and at extreme levels may be diagnosed as having what used to be called multiple personality disorder (and is now called dissociative identity disorder).

The box of the boxplot shows the IQR. The vertical line inside the box shows where the median is (recall that the median divides the upper and lower halves of the data). Because the median is not in the centre of the box we know that the distribution is not symmetric. This is clear also in the histogram. The lines that come out from each side of the box are the 'whiskers.' These whiskers are based on the size of the box (see Box 3.1). Outside of these are outliers. The boxplot shows that while most people have low scores on the DES II, a few have much higher scores. This is called being positively skewed.

I find boxplots most useful when comparing groups. Suppose a developmental psychologist asked 20 children from each of five different age groups to solve some mathematics problems. Suppose there were 20 multiple choice questions and each question had four options. The data might look like that depicted in Figure 3.2. If people were guessing randomly, they would probably get somewhere near five correct answers (this level is shown with a dashed line in the figure).[2] Most 4 and 5 year olds are at about this level, so clearly have not mastered whichever mathematical techniques are involved. The 7 and 8 year olds get most of the questions right. They are at the *ceiling*, meaning they cannot do much better. The scores on these mathematics problems differentiate among the 6 year old children. Some do poorly, some do well, and most are in the middle. If an educational psychologist wanted a test to discriminate among children, these problems would be most useful with the 6 year olds.

The Variance and Standard Deviation

The two most popular statistics to describe the spread of a distribution are the *variance* and the *standard deviation*. I will begin with the variance and then describe how to calculate the standard deviation from this.

When one thinks about how spread out a distribution is, one way to think of it is how far away most of the points are from the mean. If they are all close to the mean, then the distribution is not spread out and we would want a measure of dispersion to reflect this. If the points are far from the mean then we would want

---

2 Because there are four choices, if someone is guessing randomly, there is a 25% probability that they will answer a question correctly; 25% of 20 questions is 5.

a measure of dispersion to be large. The variance of a variable can be calculated in four steps, providing that you have already calculated the mean.[3]

1   Subtract the mean of the variable from *each* value, $(x_i - \bar{x})$.
2   Multiply this value by itself; in other words, square it, $(x_i - \bar{x})^2$.
3   Add these values together, $\sum(x_i - \bar{x})^2$.
4   Divide by the number of cases minus one, $\sum(x_i - \bar{x})^2/(n - 1)$.

The following shows the equation in full:

$$\text{var } x_i = \frac{\sum (x_i - \bar{x})^2}{n - 1}$$

It can be rather time consuming to calculate the variance for a hundred data points, but consider the following example with just 10 data points. There is much interest in how people's ability on various physical and mental tasks declines as they get older. Here 10 young adults and 10 much older adults were asked a series of current affairs questions and the times taken to complete these were recorded. The scores, rounded to the nearest second, were:

Younger	4 3 4 5 3 7 4 6 6 5
Older	5 1 9 6 4 4 10 7 4 10

Table 3.1 shows some of the calculations of the variance for the younger people. Notice that the sum of $(x_i - \bar{x})$, called the sum of the residuals, is equal to zero (all the negative values are counterbalanced by the positive ones). This is because of the way that the mean is calculated. Notice also that when each residual is squared it is positive. To get the variance we divide the sum of squared residuals (16.10) by the number of cases minus one (10 − 1 = 9) and get 16.10/9 = 1.79.

In the physical sciences, much care is taken with the units of measurement. For example, physicists are very careful differentiating 9 metres per second, which describes a velocity, from 9 metres per second squared, which describes a *change* in velocity (an acceleration). Social scientists are often less careful about this, but it can be important. Consider the data in Table 3.1. The value for x_1 is 4 seconds. The residual is − 0.7 seconds; adding or subtracting does not change the units (and you

---

3 It is probably worth restating what each of the symbols used below represents to avoid unnecessary confusion. The symbols are as follows: x_i is the value of x for the ith case (so x_3 is the score for the third case), $\bar{x}$ is the mean of x_i, $\sum$ means add together, n is the sample size, and var x_i is the variance of x_i.

Table 3.1 **Calculating the residuals ($x_i - \bar{x}$), the squared residuals ($x_i - \bar{x}$)^2 and the sum of the squared residuals $\Sigma(x_i - \bar{x})^2$, which is often just called the sum of squares. To find the variance divide by $n - 1$ (here $10 - 1 = 9$), and you get 1.79.**

	$(x_i - \bar{x}) =$	$(x_i - \bar{x})^2 =$
	$(4 - 4.7) = -0.7$	0.49
	$(3 - 4.7) = -1.7$	2.89
	$(4 - 4.7) = -0.7$	0.49
	$(5 - 4.7) = 0.3$	0.09
	$(3 - 4.7) = -1.7$	2.89
	$(7 - 4.7) = 2.3$	5.29
	$(4 - 4.7) = -0.7$	0.49
	$(6 - 4.7) = 1.3$	1.69
	$(6 - 4.7) = 1.3$	1.69
	$(5 - 4.7) = 0.3$	0.09
Sum (Σ)	0	16.10

cannot add or subtract items that do not have the same units). When a value is squared, so are the units (items can be multiplied and divided even if they do not have the same units, as in kilometres travelled being divided by hours taken to achieve kilometres per hour). The squared residual of the first person is thus 0.49 seconds squared. Summing all these values does not change the units, so the sum of the squared residuals is 16.10 seconds squared. Dividing by $n - 1$ (9) gives a variance of 1.79 seconds squared. The variance is a kind of average squared residual.

If people are thinking about the dispersion of response times, it is difficult to think in terms of seconds squared. A measure that is closely related to the variance is the *standard deviation*. The standard deviation, denoted *sd*, is simply the square root of the variance.[4] By taking the square root the units return to seconds.

$$sd = \sqrt{\operatorname{var} x_i} = \sqrt{\frac{\sum(x_i - \bar{x})^2}{n - 1}}$$

So for the above data, the variance is 1.79 seconds squared and the standard deviation is 1.34 seconds. There is another equation that is often used to calculate the variance and standard deviation. I think that most of you would like as few

---

4 The square root is the opposite of squaring. For example, if we square 3 seconds we get 9 seconds squared (3 seconds × 3 seconds = 9 seconds squared). So, the square root of 9 seconds squared is 3 seconds.

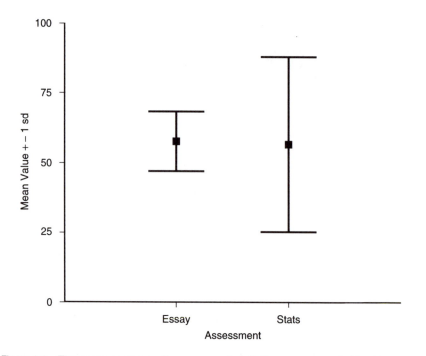

Figure 3.3   **The mean scores on the essay and statistics assessments. The bars show one standard deviation either side of the mean. The statistics assessment has the large standard deviation; the scores are more spread out.**

equations as possible, but because you may run across this, I feel that I should provide it:

$$sd = \sqrt{\operatorname{var} x_i} = \sqrt{\frac{\sum x_i^2 - (\sum x_i)^2/n}{n-1}}$$

For this equation you do not need to know the mean before doing the calculations. Lots of textbooks say this is a big advantage. However, in almost every case you would also want to calculate the mean. Further, the meaning of standard deviation is less transparent from this equation, and so less conceptually useful.

Standard deviations can be placed onto graphs depicting means to show additional information. This extra information is about the spread of the data. The data from the variables, ESSAY and STATS, are shown in Figure 3.3. The standard deviation for ESSAY is 10.86 and for STATS is 31.65. This graph was done using SPSS. It uses the mean as the measure of central tendency and the standard deviation as the measure of spread. This graph can be made without a computer in four steps.

1   Draw a point where the mean is for each variable.
2   Subtract the standard deviation of each group from that group's mean. Draw a short horizontal line at this point below the dot. This becomes the lower bound.
3   Add the appropriate standard deviation to each mean and draw a line above the point to make the upper bound.
4   Join these two bounds with a vertical line. The point for the mean should be exactly in the middle of this line.

This graph shows that the scores for the statistics-based exam are more spread out.

Summary

The goal of this chapter was to introduce you to some measures of dispersion. Dispersion is important for two main reasons. First, in its own right, in that if one variable or group has a much larger spread than another this may be of importance. Second, the spread of a variable relates to how confident we can be about our estimates of central tendency. It is because of this that measures of spread are used when assessing the precision of statistics. This will be discussed further in the second half of the book.

The measures that were discussed were the range, the interquartile range, the variance and the standard deviation. The range is simply the difference between the highest and the lowest points. The interquartile range shows where the middle 50% of the sample lies. The variance and standard deviation, which are closely related (the square of the standard deviation is the variance, or alternatively the square root of the variance is the standard deviation), are measures of how far away the points tend to be from the mean.

A logical question is: why learn different measures of dispersion? The reason is that each is useful in different circumstances. Throughout this book several alternative methods are discussed. Each asks a slightly different question of the data and provides a different answer.

Exercises

3.1   Find the IQR (interquartile range) of the statistics exam scores from Table 2.1. How does this differ from the IQR of the essay exam? How does this differ from the normal range calculated above?

3.2   Make a graph that contains boxplots for both the statistics and the essay exams.

Table 3.2 **Median severity ratings for 15 cases.**

Case	Before–severity	After–severity
Reynolds	5.5	6.0
Glover	5.0	5.0
Lawson	4.3	4.5
Williams	5.0	5.0
Smith	5.5	6.0
Nelson	5.0	5.0
Hughes	5.0	5.0
West	4.5	5.0
Douglas	4.0	4.0
Crandall	4.0	4.0
Sanders	3.5	3.0
Windsor	3.0	2.0
Stanley	1.0	1.5
Dulworth	0.3	0.0
Newton	0.0	0.0
Means	3.71	3.73

3.3 Calculate the variance for the older group (remember what you have to calculate first) of the response time data presented earlier this chapter. How does this compare with the variance of the younger group (see Table 3.1)? Which set of scores is more 'dispersed'? Why might this be the case?

3.4 Reeve and Aggleton (1998) were interested in people's memories for '*The Archers*,' a popular UK radio soap opera. Participants ($n = 48$, 12 in each of four conditions) were given one of two made-up scripts. One was of a typical day, a visit to a livestock market. The other was of an atypical event (for the Archers), a visit to a boat show. Participants were either Archer novices or experts and were asked 22 questions about the script they were given. The means and standard deviations for the experts were 15.08 ($sd = 4.17$) for the livestock market, 8.42 ($sd = 2.11$) for the boat show, and were 9.50 ($sd = 4.48$) and 8.42 ($sd = 3.23$), respectively, for the novices. Make an appropriate graph to display these results and comment on what they show.

3.5 In Chapters 1 and 2 you were presented with rainfall in centimetres for two hypothetical places for a week. Here are these data:

	Sun	Mon	Tue	Wed	Thur	Fri	Sat
Place 1	0	0	1	6	0	0	0
Place 2	1	1	1	1	1	1	1

What are the standard deviation and variance of the rainfall for each of these two places?

3.6. Schkade et al. (2000) collected data from jurors on 15 different cases. They ask for ratings of severity of the crime, on a 0 (no punishment) to 8 (extremely severe punishment) before jury deliberation and after jury deliberation. The medians for each case are as given in Table 3.2.

Find the standard deviation and variances of these values. Speculate why any difference in standard deviation between before and after deliberation ratings might have occurred.

3.7 At the beginning of this book it was stated that 'it takes *about* twenty minutes to cook rice' is a statistical phrase. Suppose that you were hired to come up with the time to write on the side of the containers for spicy cheese bread and a vegetable stir fry. With various different cooking arrangements in different climates and at different altitudes, you cooked these products to perfection. Here are the times, in minutes:

| Spicy cheese bread | 8 | 15 | 12 | 16 | 11 | 8 | 14 | 9 | 12 | 15 |
| Veggie stir fry | 11 | 14 | 10 | 12 | 12 | 13 | 11 | 13 | 13 | 11 |

Which dish would you give the longer time to, and why? Think about the purpose of the estimate before giving your answer. The correct answer is not that they should have the same estimated cooking time.

Further Reading

Kirk, R. E. (1999). *Statistics: An introduction*. London: Harcourt Brace.
    Chapter 4 of this textbook goes through each of the measures discussed here, in slightly more detail.

4 | Sampling and Allocation

The previous three chapters dealt with what are often called *descriptive* statistics. Included in these are the measures of central tendency and dispersion which are important in describing the distribution of a variable. In addition, the importance of graphing the data to provide more information about the distributions was stressed. These procedures provide the core for doing most of the statistics in this book. From Chapter 5 onward these statistics will be used to build *inferential* statistics. Inferential statistics are used when you are trying to infer something about the entire population of interest from just the sample used in your study. How this sample is gathered is therefore vital to whether this inference can be done.

Because this is a statistics book, rather than a methodology book, the focus is mostly on statistical techniques. However, one topic usually covered in methodology books is so vital for statistics that I am going to take a quick detour from purely statistical issues and look at this methodology topic: sampling participants and allocating them to conditions. I have opted to write this chapter in a more 'methods style' than a 'statistical style'. This means it focuses more on conceptual issues and – I hope that none of you mind – there are no equations. After three chapters with several equations, and even more to follow in later chapters, I think most of you will not mind a break.

Golden Standards

Most methodology and statistics books lay down firm rules on exactly how sampling and random allocation should be carried out. Statistical tests assume

Figure 4.1   **Sampling is done to select a subset of the population to be in the sample. In experimental research this sample is then allocated into conditions.**

that these are done in particular ways. While I describe these 'gold standards' of sampling and allocation, I also note that they are often not used even in very good research. I will discuss when certain aspects of these rules are critical and when they become less critical.

Besides assisting in your own endeavours, a good knowledge of sampling/ allocation issues is a great asset for evaluating the work of others. Often people describe their research as if they have used the 'gold standard' when in fact they have not. This allows the astute reader to question whether the deficiencies are detrimental to the study or not.

Sampling and allocation can be divided into two stages in the typical study (see Figure 4.1): first, deciding on a sample from the population and then, if appropriate, allocating people from this sample to the different conditions. In non-experimental or correlational studies, the final step is not used. I will discuss these stages separately and refer to the first as *sampling* and the second as *allocation*.

A few preliminary definitions are worth making here (some are repeated from Chapter 1). First, a *population* is the entire set of people or items (or anything) for which the researcher wants her/his study to be applicable. When you investigate a research question, you have in mind a population for which your results should hold. Sometimes you would want your results to hold for all humans or all inhabi- tants of a country, but sometimes the population is smaller. For example, you might only be interested in people with Korsakoff's syndrome or members of a particular militia.

Sampling is the process by which people are chosen from a population. For simplicity I will assume people are the cases being sampled unless otherwise stated (for another type of case below, I sample pizza toppings). The *sample* is the resulting set of people chosen, the outcome of the sampling procedure. *Allocation* is the process by which people in the sample are allocated to conditions. Sampling and allocating are the processes; the sample and the conditions are the outcomes. The hope is usually that the sample is *representative* of the population and that the people in each condition are representative of the sample. Representative means that the characteristics of the sample are similar to those of the population. Thus, you would not want your sampling procedure to produce a sample of people all over 6 feet (1.82 m) tall if the mean height in the population of interest was 5 feet 6 inches (1.68 m). Similarly, if your sample was about half males and half females, you would not want one of the conditions to be all of one gender. Good sampling and allocation procedures reduce the likelihood that the sample is unrepresentative of the population and that the conditions are unrepresentative of the sample.

Sampling

The 'Gold Standard' for Sampling

The 'gold standard' is that the participants are sampled at *random* from the population. The word 'random' is critical here. The English dictionary definition of 'at random' is usually something like 'without aim or purpose or principle' (Allen, 1985: 613). In any science, certain terms which have perfectly adequate definitions for everyday life are given more precise meanings. In statistics and methodology, random means that each possible sample is equally likely to be chosen. Sometimes this is referred to as a *simple random sample* (SRS). As most statistical techniques assume that this has been done, I will describe exactly what it is, why it is an important assumption for the statistical techniques used in behavioural sciences, what alternatives exist, and what possible problems there are with the alternatives.

First, I will describe in more detail what an SRS is and what it is not. It will be easiest to do this by using an example with a small population, say five items. These items might be your siblings, your lecturers, or many other things. When I go with my methodologist friends to the pub, conversation seems always to lead to pizza (and sampling theory).

My local pizza outlet has five possible toppings: mushrooms, peppers, olives, sausage and pepperoni. So, the population is the five toppings. It also has a special price for large pizzas with any two toppings. Because methodologists are frugal, and we carefully calculated that this was a good buy, we always get this. Our sampling procedure involves choosing two toppings. If we were going to take an SRS then any of the 10 combinations shown in the first column of Table 4.1 would be equally likely to be chosen. Since there are 10 possible combinations, each has a probability of 1 in 10, or a 10% chance, of being chosen if we sampled at random.

The word *probability* is tricky. There are entire books written about what it means and even experts disagree. Here I have used it to mean that in the long run, after hundreds or thousands of trials (yum yum!), each one of those 10 possible combinations would be ordered approximately one-tenth of the time.

The important aspect of an SRS is that it tells us some likely characteristics about the samples. We would expect mushrooms to be on the pizza about 40% of the time. Also, we would expect a vegetarian pizza about 30% of the time. This is because, of the 10 possible pizzas, three would be suitable for vegetarians (assuming vegetarian cheese is used). The word *about* is also important. If

Table 4.1 **Pizza combinations.** The first column shows all possible pizza combinations. The second is where all toppings have an equal probability of being picked, but random sampling from these pizzas will not be a simple random sample of the toppings. The third column shows the possible samples for vegetarian pizzas (there would be only three toppings in the population). The final column lists the possible samples from a quota sample where one of the toppings has to be meat. It is worth noting that the order of the toppings is irrelevant: mushrooms & peppers is the same as peppers & mushrooms.

All possible combinations (equally possible with SRS)	All toppings equally likely	Vegetarian samples	One meat quota samples
Mushrooms & peppers	Mushrooms & peppers	Mushrooms & peppers	Mushrooms & pepperoni
Mushrooms & olives	Sausage & olives	Mushrooms & olives	Mushrooms & sausage
Mushrooms & pepperoni	Olives & pepperoni	Peppers & olives	Peppers & pepperoni
Mushrooms & sausage	Pepperoni & peppers		Peppers & sausage
Peppers & olives	Sausage & mushrooms		Olives & pepperoni
Peppers & pepperoni			Olives & sausage
Peppers & sausage			
Olives & pepperoni			
Olives & sausage			
Pepperoni & sausage			

1000 pizzas with two toppings were ordered (i.e. if 1000 SRS samples of size 2 were taken), we would not expect exactly 300 (30%) to be vegetarian pizzas. Statistics work because we know how unlikely it is, assuming an SRS is used, that the sample will have certain characteristics. For example, if more than 500 of the pizzas were vegetarian, then it would be very unlikely that an SRS was not being used. We would conclude that there was a non-random process in the sampling, presumably caused by one of the methodologists being a vegetarian.

There are some common misconceptions about what constitutes an SRS. The most common one is still using the dictionary definition of the word random, meaning that the toppings were chosen in some haphazard manner. Another common misconception is that having some chance element in the sampling makes it an SRS. In the pizza example, suppose a coin was flipped to determine whether to have mushrooms or not and then the choice was random. This is not an SRS because there will be a 50% probability of having a mushroom pizza, and we know that it should only be 40% with a simple random sample. Finally, some people think that the SRS is defined by each element (i.e. topping) having an equal chance of being chosen. As there are five toppings and two toppings picked, having equal probability would mean each topping was chosen about 40% of the time. The SRS has this characteristic, but so do lots of other forms of sampling. In the second column of Table 4.1 are five pizzas. If we randomly sampled from this group of five, each topping would be chosen about 40% of the time. However, not all combinations are possible and this can make some differences. For example, the probability of a vegetarian pizza is only 20% if taking a random sample from this group, while it is only 30% from an SRS.

While an SRS is rare, there are a few very well-known examples of it. One of these is the UK's National Lottery (and most lotteries in the United States and other countries). The population is the 49 balls and six are sampled (ignoring the 'bonus ball'). Any possible combination of the six balls is possible. There are about 14 million possible combinations so the probability of winning the jackpot with one lottery ticket is about 1 in 14 million. While the lottery is random (though see Box 4.1), we can calculate the probability of the sample of six balls having certain characteristics. For example, it is known that having all six balls with even numbers on them should happen slightly less than 1% of the time.[1]

BOX 4.1   WHEN RANDOM SAMPLING ISN'T

In most research small deviations from random sampling are unlikely to be important. But what if it could cost you your life?

At several points throughout US history young men have been conscripted into the military based on a lottery system. Fienberg (1971), in an excellent article describing what 'random' means, discusses some of the problems with the 1970 lottery which determined if men were sent to Vietnam. The lottery system was based on people's birth dates: 366 capsules, one for each possible birth date, were placed in a box. The January capsules were put in and pushed over to one side. Next the February capsules were put in, and pushed

---

1 Here is a tricky question. The probability of having all six balls with odd numbers on them is slightly *greater* than 1%. Why the difference? (Don't try working out the numbers. If you think you know the reason you are probably right.)

up next to the January ones. This continued for the other months, and resulted in capsules for dates early in the year (i.e. January, February, etc.) being predominantly on one side and capsules for later dates on the other. The box was shaken, carried around a bit, and then its contents were poured into a bowl. This resulted in the early months capsules being towards the bottom of the bowl, and the late months capsules being towards the top. The capsules were then picked. According to Fienberg's sources, they were generally picked from the top of the bowl. The result was that people were much more likely to have the ball with their birthday on it picked first if they were born in the late months and therefore be drafted first.

It is worth mentioning that the draft was done differently the following year.

Alternatives to an SRS

There are several alternatives to an SRS. I will describe three of them: cluster sampling, quota sampling and convenience sampling.

A *cluster sample* is where the researcher first samples large clusters of people, like whole neighbourhoods or districts, and then samples people within these clusters. If doing a face-to-face survey, this means the interviewer can go to one designated location and sample several people from there, and then move on to another. An example that could be used in educational research is shown in Figure 4.2. The population may be all pupils in schools in the United Kingdom. If an SRS sample was taken, the researcher might have to travel to several different schools to test only one or two pupils in each. This would be a waste of effort and money. Instead, the researcher might take an SRS of schools. A handful of schools would be sampled, and then within these schools (or clusters) individual pupils could be sampled. Figure 4.2 is described as a two-stage cluster sample, sampling schools in the first stage and then sampling pupils. In general the estimates are less accurate than you get with an SRS. However, the cost savings are often quite large. Although there are methods to analyse clustered data, these are beyond the scope of this book (see Wright, 1998a).

Quota samples are very popular in social science. In quota sampling the interviewers or researchers choose the sample so that there are specific percentages of various groups. An example might be that they want their sample to be half women and half men. In some cases the researcher may purposefully choose the sample to be non-representative of the whole population. With the pizza topping example, suppose my pub companions and I wanted one meat topping and one vegetable topping (see Table 4.1, fourth column). This is a type of quota sample but it overrepresents the meat toppings (both meat toppings have a 50% chance of being sampled while each vegetable has a 33% chance). Consider another example. If a researcher was comparing left- and right-handed people, s/he would probably want approximately equal numbers of people in these

Stage 1

Stage 2

Sample some of
the schools in
the district

Sample some of
the pupils from
each of the schools
sampled in the
first stage

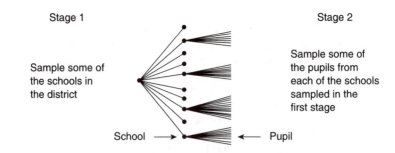

School ⟶ ◄ Pupil

Figure 4.2 **A two-stage cluster sample of schools and pupils.**

groups, even though about 90% of humans are right handed. Thus, while an SRS would probably produce a sample that was about 90% right handed, a researcher using a quota sample could specify that s/he wanted left-handed people to be oversampled.

The final technique produces what is called a *convenience sample*. As the name suggests, this sample is the easiest to attain. This is the most common sample in psychology and in many cases it is justified. Convenience samples, which are sometimes called opportunity samples, include where you just go around and ask the first 20 people you can find in the school cafeteria, or use the first group of people who sign up for your study. The problem with convenience samples is that it is difficult to justify generalising your findings to the population at large.

Allocation

The 'Gold Standard' for Allocation

As with sampling, there is a 'gold standard' for allocating people from these samples to the experimental conditions. This is called *random allocation* and means each person is randomly allocated to a condition. For example, if there were two conditions, you could flip a coin for each person to determine which condition the person is in.[2] Random allocation is much easier to do than an SRS, and so it is often used. There are some slight variations from it. For example, some people will put the first person into condition 1, the second into condition 2,

---

2 I know that coins are not perfectly balanced, but realistically most are close enough.

the third into condition 1, etc. This is not random allocation, but it is not a bad approach and ensures there are approximately the same number of people in each condition. What would be bad, for example, is if all the people sampled in the morning were in one group, but those sampled in the afternoon were in another. This would introduce a *bias* into the allocation. For example, it might mean that only people who wake up early are in the first group. Another bias would be choosing people whose names are at the beginning of the alphabet to be in one group and those with names at the end to be in the other. It could happen that there are systematic differences between these groups. Box 4.2 describes biased allocation in a large-scale study.

BOX 4.2 WHEN RANDOM ALLOCATION ISN'T

In 1930 an experiment with 20,000 children was carried out in Lanarkshire, Scotland, to test how giving children milk every day affected height and weight. Children's height and weight were measured at both the beginning and end of the experiment. William Gossett, who went by the pseudonym 'Student' (1931) and who you will hear more about in Chapter 6, described a variety of problems with the design of this study. Here I describe just one: that the allocation of students to conditions was made by the teachers. The teachers were supposed to allocate children randomly to the control group, who received no milk, and to the experimental group, who received milk.

According to a report 'Student' quotes from, teachers were allowed to substitute well-fed or ill-nourished children if the control and experimental groups in their classrooms did not appear equally nourished. This allowed the teachers to bias the allocation. 'Student' states:

> it would seem probable that the teachers, swayed by the very human feeling that the poorer children needed the milk more than the comparatively well to do, must have unconsciously made too large a substitution of the ill-nourished among the 'feeders' and too few among the 'controls' (1931: 399).

When the initial heights and weights were compared the 'control' children were taller and weighed more (see his Diagrams 1–4). It is likely, therefore, that the controls differed in several other ways, including general health and socio-economic status. 'Student' summarised this expensive (£7500 in 1930, which is about £300,000 now) study:

> though planned on the grand scale, organised in a thoroughly business-like manner and carried through with the devoted assistance of a large team of teachers, nurses and doctors, [it] failed to produce a valid estimate of the advantage of giving milk to children ('Student', 1931: 406).

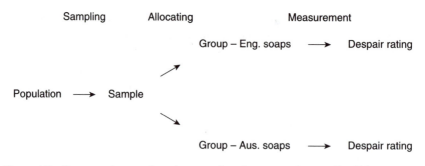

Figure 4.3   **An example experiment comparing viewer reactions to English and Australian soap operas.**

Random allocation is extremely important in interpreting experimental results. Figure 4.3 shows a typical experiment. In this there is a population from which a sample is chosen, and then people in this sample are allocated to one of two conditions. The conditions are treated differently. Let's say the experiment is on whether people enjoy English or Australian soap operas more. The first group were forced to watch 10 hours of an English soap opera and then physiological instruments rated their despair. The second group had the same measures taken, but after watching 10 hours of an Australian soap. The researcher would probably take the mean despair rating for each group and see which is larger. Now people differ in how much they dislike soaps. If the two samples differ greatly on this, then any difference on the final despair measure could reflect this. Therefore, in an ideal circumstance, the two conditions would be exactly the same on overall soap dislike prior to study. While random allocation does not guarantee this, as with random sampling, we do know how unlikely it is to have two groups which differ greatly.

It is easier to allocate people randomly to conditions than to have an SRS of a large population. There are several ways to do random allocation and it is (usually) not any more expensive than other methods. Because of this it is expected that random allocation is used in experiments.

Summary

The most elegant statistics, the most informative graphs and the clearest of writing styles are all for naught if the design of the study is poor. Depending on the purpose of the study, if poor sampling or poor allocation is used, the study may be worthless. When trying to estimate the value of some variable for a population, it is necessary to have the sample representative of the population. If not, whatever oddities exist in the sample may be errantly inferred to exist in the population.

When doing experiments, the critical aspect is how people in the sample are allocated to conditions. Before the manipulation, the people in the different conditions should be as similar to each other as possible. If they are markedly different, this difference may lead to incorrect inferences about the effect of any manipulation.

The gold standards of sampling and allocation are simple random sampling and random allocation. These standards are assumed in most statistical tests. Yet, particularly with reference to sampling, often other methods are used. While gross violations, like that described in Box 4.1, mentioned earlier, are unacceptable, small variations are usually acceptable. What this usually means is that your statistics are not as precise as they should be. Many statistics textbooks take a hard, and unproductive, line. They state that all variations are unacceptable and that the data should be either thrown out or analysed in much more complex ways. This is not helpful. Much useful data do exist where these standards are not met. The main point I want to stress is that the gold standards are what you should aim for, but if you fall short then the sample should be as representative of the population as possible, and the conditions as similar (pre-manipulation) to each other as possible. In practice, when the gold standards are not met, the statistics calculated are less precise than they appear, so you should be extra cautious.

Exercises

These exercises are not statistics exercises, but methods questions. Ask your instructor for the appropriate length for your answers. For each, describe the population, the sample and any experimental conditions. For each also assume that you are on a limited budget.

4.1 Design a study to investigate the effects of television violence and children's aggressive behaviour.

4.2 An environmental scientist feels that the UK people's attitudes towards the problems with the rapidly increasing global population are changing. Design a study to help to answer this question.

4.3 A marketing firm has just thought up a new advertising campaign to use for its toothpaste. Design a study to evaluate this campaign.

4.4 Around half of the time that someone is falsely convicted of crime, the main evidence against them is errant eyewitness testimony. A psycho-legal researcher wondered if having witnesses first try to draw the person helped them in later attempts to identify the culprit in an identification parade (i.e. a line-up). How would you go about testing this hypothesis?

4.5 In the United Kingdom, the party in government can call when it wants to have an election. Obviously it only wants to call an election when it thinks that it will win. Suppose that you were asked by the party leader to design a study to see whether s/he should call an election. How would you do this?

4.6 What is a simple random sample?

4.7 Define 'probability.'

Further Reading

Cook, T. D., & Campbell, D. T. (1979). *Quasi-experimentation: Design & analysis issues for field settings.* London: Houghton Mifflin.

   This is one of the best books ever on social science methods. Chapter 1 provides an excellent discussion on causation.

Fienberg, S. E. (1971). Randomization and social affairs: The 1970 draft lottery. *Science, 171,* 255–261.

   People were sent to war and died because of biased sampling. This is the best paper I have read on sampling.

Wright, D. B. (1998b). People, materials and situations. In J. A. Nunn (Ed.), *Laboratory psychology* (pp. 97–116). Hove: Lawrence Erlbaum.

   In this chapter I discuss how we do not just sample people, but also sample the materials we use in our research and the context in which our research is carried out.

5	Inference and Confidence Intervals

The main purpose of statistics is to use information about a sample to infer something about a population. This is called *inference*. It involves using characteristics of the sample to make some best guesses for characteristics about the population. In this chapter I describe how this is done for the mean of a variable and for the difference between two means. You will learn what confidence intervals are and how to construct them.

Inferring a Population Mean: Constructing Confidence Intervals

In one sense, the inference from a sample mean to a population is simple. If we have used an SRS, then a sensible guess for the population mean is just the sample mean. If an SRS of 100 people found the mean number of cigarettes smoked a day was 6.2 cigarettes, then this is a good estimate for the population mean. The lower case Greek letter μ (pronounced *mu*) is used to refer to population mean. We never know exactly what the population mean is, unless we sample the entire population. The purpose of statistics is to get the best estimate.

While it is useful knowing that the best single-point estimate for a population mean is the sample mean, in most cases you want to give an interval which you feel is likely to contain the population mean. This is more complicated. When a newspaper article states that the average person consumes 42 ± 2 grams of broccoli a year (the $\pm$ means 'plus or minus'), it means that, based on some survey, they have some level of confidence that the population mean for grams of broccoli eaten in a year is between 40 and 44 grams. The width of a confidence interval tells you how precise your estimate is. If the newspaper reported the amount was 42 ± 30 grams, then you would only be confident that the population mean was somewhere between 12 and 72 grams. This latter interval is not very informative because it is so imprecise.

To calculate a confidence interval you need to know the number of people in the sample, the sample mean and the sample standard deviation. You also have to use a table of a statistical distribution and decide on what confidence level you want to use. I will go through an example to show how confidence intervals are made.

Newton (1998) was interested in hostility levels on arrival at and at discharge from Grendon Prison in the United Kingdom. She gave a sample of 94 people the *Hostility and Direction of Hostility Questionnaire* (HDHQ) at both time points. This questionnaire produces a total hostility score (high scores mean more hostility). According to Caine et al. (1967) the mean in the general population is about 13.0. The mean score on arrival for the 94 prisoners was 28.3 with a standard deviation of 8.0.

The equation for the 95% confidence interval is

$$CI_{95\%} = \bar{x} \pm t_{0.05} \frac{sd}{\sqrt{n}}$$

There is a new symbol here that you have not encountered.[1] It relates to the t distribution which is located in Appendix C. It is fairly complex to calculate the value of $t_{0.05}$ but it is usually around 2 (Appendix C gives further details on how to use the table). The letters df in the table stand for *degrees of freedom*. When calculating the confidence interval of a single variable the degrees of freedom are $n - 1$. With 94 prisoners in Newton's data set, $df = 93$. In Appendix C you should go to the row with 93 in the df column and look for the value in the $t_{0.05}$ column. However, there is no row corresponding to $df = 93$. You should go to the closest row, here $df = 90$. The value is 1.99. The '0.05' in $t_{0.05}$ is due to the 95% in '95% confidence interval'. It is because 100% – 95% is 5%, which is 0.05. The reason why 5% is printed in the table instead of 95% will become clear when discussing hypothesis testing in Chapter 6. Inserting these values into the equation yields

$$CI_{95\%} = 28.3 \pm 1.99 \frac{8.0}{\sqrt{94}} = 28.3 \pm 1.6$$

---

1 Those which you have already encountered are for the mean ($\bar{x}$), the standard deviation (sd) and the sample size (n).

Sometimes the 1.6 is added to or subtracted from the mean and the confidence interval is written as (26.7, 29.9).

So what does having $CI_{95\%}$ of 28.3 ± 1.6 mean? We expect that about 95% of the time when a confidence interval is made that the population mean (μ) will be within the interval. This allows us to be fairly confident that the confidence interval we calculate contains the population mean (see Box 5.1).

BOX 5.1 CONFIDENCE INTERVALS AND DEAD CATS

The view of most methodologists is that if you have just found a 95% confidence interval, it does *not* mean that there is a 95% probability that the true mean lies in *that* interval. They argue that the population mean, which is denoted with the Greek letter μ (mu), either is or is not in the observed interval.

This situation is very similar to a famous physics experiment called Schrödinger's cat experiment (Schrödinger, 1983, originally published 1935). This experiment was prompted by some disagreements in quantum physics and Schrödinger wanted to convince people that some aspects of probability states in the microscopic world of quantum physics do not manifest themselves in our macroscopic world. Schrödinger placed a cat in a box, with a radioactive substance, a Geiger counter, a hammer and a flask of poison. There is some probability associated with an atom of the substance decaying. For argument's sake, let's say that the probability is 95% that an atom will decay in an hour. If it does, this will be measured with the Geiger counter, which triggers the hammer to smash the flask, releasing the poison and killing the cat. There is no way to look into the box. The question is: if you cannot look into the box, what is the cat's condition after being in this box for one hour?

One interpretation is that, at least at the microscopic level of quantum physics, the cat can be both alive and dead. One might say that, without looking in the box, the cat was 95% dead. When the box is opened, the cat either makes an immediate and miraculous recovery, or instantaneously dies. However, for most people when considering the macroscopic level of cats, the cat is either alive or dead. This was what Schrödinger was trying to show, that quantum 'weirdness' does not manifest itself in the macroscopic world, and many people agreed. Einstein, for example, stated that 'nobody really doubts that the presence or absence of the cat is something independent of the act of observation' (Przibram, 1967: 39).

In the same way, when you calculate a 95% confidence interval, this particular interval either includes the population mean, or it does not. We do not know which because we do not know the true mean; we cannot open the box. What we can say is that if you repeat, over and over, calculating hundreds of confidence intervals, then approximately 95% of the time this procedure will produce an interval that will include the population mean. But the mean is either inside or outside of the interval, it is not part inside and part outside.

Now because this sounds pretty close to '95% probability that *the* confidence interval includes the population mean', it is generally agreed that people are allowed to say that they are '95% confident that *the* confidence interval includes the population mean'. This is not an ideal solution, but it has become a convention. It is simply because the word 'confident' carries less philosophical baggage than 'probability.'

I should probably add that the traditional view of Schrödinger's cat experiment was that it was just a *thought experiment*. I felt comfort thinking this because if it were a real experiment there might have been ethical considerations. Imagine my dismay when reading a webpage from the High Power Laser Group at the University of Oxford (http://users.ox.ac.uk/~jsw/Schroedinger.html in February 2001) which claimed that Schrödinger had a 'pathological hatred of any small, cuddly animal that meowed' and that he actually performed this macabre experiment. This group's webpage also describes some of their own replications, using a Hibachi grill and local Oxford cats. From their webpage you can take part in these replications. Like the debates about the meaning of probability, the debate about whether the experiment actually took place, between the traditionalists and the High Power Laser Group, will remain heated.

Figure 5.1 shows the *t* distribution with 93 degrees of freedom ($df = 93$). The area underneath the curve can be thought of as the total probability of some event. If we want to include 95% of this probability in our confidence interval, this means we exclude 2.5% at each extreme. The shaded areas in Figure 5.1 contain, in total, 5% of the total area under the graph. Statisticians say that there is 2.5% in each *tail*.

If we wanted a 99% confidence interval we would want to exclude a total of 1% of the area under the curve. This means only 0.5% of the total area under the curve would be in each tail. Here are the numbers for Newton's 'at arrival' data for a 99% confidence interval:

$$CI_{99\%} = \bar{x} \pm t_{0.01} \frac{sd}{\sqrt{n}}$$

$$= 28.3 \pm 2.63 \frac{8.0}{\sqrt{94}} = 28.3 \pm 2.2$$

Notice that this new confidence interval is larger, going from 26.1 to 30.5, than the 95% confidence interval. By increasing the confidence you have in the population mean being within the interval, the interval becomes larger. It is a tradeoff between confidence and precision. The most common confidence intervals are the 95% and 99% ones and usually the 95% confidence interval is used. It is fairly

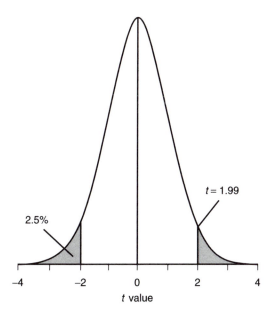

Figure 5.1 **A *t* distribution with 93 degrees of freedom. The shaded area, where *t* < – 1.99 or *t* > 1.99, includes 5% of the total area under the curve.**

arbitrary that this particular level is used, but because it has become a convention it will be used here.

When constructing and interpreting a confidence interval, three assumptions are made. The first is that simple random sampling was used. In this case we would probably assume the population was all the prisoners who might be in a unit like Grendon. Newton (1998) describes the sampling and notes that some prisoners who are only there a short time could not be part of the sample. Thus, it is important to make sure that the inference is not made to all people or even all prisoners.

The second assumption is that the data points are independent. This means that having one person sampled should not determine whether someone else is chosen. It also means that the responses of one person should not influence what someone else says. So, for example, with a hostility questionnaire it would be bad to test people in groups because if there is an extremely hostile person present, this might affect everybody's scores.

Finally, the distribution in the population is assumed to be normally distributed. The normal distribution was described in Chapter 2 and is shaped very similarly to Figure 5.1. The assumption of many statistical tests is that the population distribution is normal. Since researchers seldom have data from the entire population, they usually look at the sample distribution and see if it looks roughly normally distributed. If it does, then this assumption is usually made.

In summary, confidence intervals are very useful: they provide information about an estimate of the population mean and also how precise the estimate is.

Examining the Difference between Two Means for the Same Person

Constructing a Confidence Interval for the Difference

Social scientists often differentiate 'between-subjects' and 'within-subject' studies. Between-subjects studies are where you are comparing different groups of people. Within-subject studies are where you are comparing one group of people in different situations. In between-subjects studies each person is in only one condition, while in within-subject studies each person is in multiple conditions. In this section I will describe a simple within-subject design, where there are only two conditions and each person takes part in both conditions. I will first describe an example on coffee preference with a small number of participants, and then examine Newton's (1998) data set, comparing the prisoners' hostility levels when they arrived at Grendon with when they were discharged.

Suppose 10 regular coffee drinkers were asked to taste two cups of coffee and to rate them for enjoyment on a 1 to 7 scale. The data are shown in Table 5.1. The first coffee they tasted was a freshly ground coffee. The second was a cheap instant coffee. Not surprisingly most people liked the more expensive fresh coffee.

The first two columns give the scores people gave for the different coffees, $FRESH_i$ for the freshly ground coffee and $INSTANT_i$ for the instant coffee. $DIFF_i$

Table 5.1 **Data from 10 participants comparing how much they like two different types of coffee.**

	$FRESH_i$	$INSTANT_i$	$DIFF_i$	$DIFF_i - \bar{x}$	$(DIFF_i - \bar{x})^2$
	5	3	2	1	1
	4	3	1	0	0
	6	5	1	0	0
	3	4	-1	-2	4
	4	4	0	-1	1
	5	3	2	1	1
	6	3	3	2	4
	3	3	0	-1	1
	5	3	2	1	1
	4	4	0	-1	1
Sum	45	35	10	0	14
Mean	4.5	3.5	1.0	0	1.56*

* This is not the actual mean. It is the variance of the variable $DIFF_i$, the sum of squares divided by the number of cases minus one (14/9). I calculated this because it can be used in later calculations.

is simply FRESH$_i$ minus INSTANT$_i$, being careful to make sure that if the person liked the instant coffee more that they have a negative difference (only the fourth person did, $3 - 4 = -1$). The next step should be to find the 95% confidence interval for this variable, DIFF$_i$, in the same way as was done with the hostility scores in the last section. The next two columns are used to help calculate the standard deviation which is used in the confidence interval equation.

The final column of Table 5.1 gives the values for $(\text{DIFF}_i - \bar{x})^2$. Recall from Chapter 3 that the variance is $\Sigma(x_i - \bar{x})^2/(n - 1)$. This is the number in Table 5.1 marked with the asterisk. If we take the square root of it we get the standard deviation: $sd = 1.25$. We already know the mean of DIFF$_i$ (1.0) and the sample size ($n = 10$). With $n = 10$, there are nine degrees of freedom. If we go to Appendix C and look up $t_{0.05}$ for $df = 9$ we find $t_{0.05} = 2.26$. Putting all these into the confidence interval equation yields

$$\text{CI}_{95\%} = 1.0 \pm 2.26\frac{1.25}{\sqrt{10}} = 1.0 \pm 0.89$$

So the 95% confidence interval goes from just a little above zero to about two. Therefore, we would have some confidence that people do like the fresh coffee better, but with only 10 people in the sample we cannot be very precise about how much more people like freshly brewed coffee.

Let's consider another example. As discussed earlier, the main purpose of Newton's research was to examine differences in hostility between when inmates entered Grendon Prison and when they were discharged. At discharge the HDHQ was also administered (the mean was 21.6 with a standard deviation of 9.2), and Newton compared these with the earlier scores. She subtracted the scores for each person and found the mean of this difference was -6.6 with a standard deviation of 9.0. We can then find the 95% confidence interval for this difference variable in the same way as we did with the coffee example.

$$\text{CI}_{95\%} = -6.6 \pm 1.99\frac{9.0}{\sqrt{94}} = -6.6 \pm 1.8$$

The confidence interval runs from -8.4 to -4.8. As this confidence interval does not overlap with zero we can say that we are confident the scores were lower at discharge. A 99% confidence interval is -6.6 ± 2.4, which also does not overlap with zero. This interval is larger, so is less precise, but we have more confidence that the 99% interval will contain the true value than we did with the 95% interval procedure.

Confidence Intervals for the Difference between Groups

Tatar (1998) sampled 295 Israeli secondary school students and asked them to rate (on a 1 to 5 scale with 5 being high) how much several descriptions characterise, in their opinions, a 'significant' teacher. One of the descriptions was 'makes me learn willingly'. In this sample, 166 were girls and 129 were boys. The mean for girls was 3.51 with a standard deviation of 0.99. The mean for boys was 3.26 with a standard deviation of 1.05. Tatar was interested in this gender difference. One approach would be to calculate the 95% confidence intervals for each gender individually. The calculations for the confidence intervals are

$$\text{for girls: } CI_{95\%} = 3.51 \pm 1.98 \frac{0.99}{\sqrt{166}} = 3.51 \pm 0.15$$

$$\text{for boys: } CI_{95\%} = 3.26 \pm 1.98 \frac{1.05}{\sqrt{129}} = 3.26 \pm 0.18$$

The degrees of freedom are 165 and 128 for the girls and the boys, respectively. When the degrees of freedom are above 100 you should use the row for $df = 100$. The 1.98 is found from the t table (in Appendix C) in the row for $df = 100$ and in the 0.05 column. So we can say with some confidence that the population mean for girls is between 3.36 and 3.66 and that the population mean for boys is between 3.08 and 3.44 (Figure 5.2). The two confidence intervals overlap. If they did not then we could be fairly sure that the population mean for girls was higher.

Although the confidence intervals overlap, this does not mean that the means are the same. In order to examine if there is a gender difference, we need to find the confidence interval for the difference. When you are comparing two variables for the same person, as in the coffee example, you can simply calculate a difference variable for each person. Here we want to compare two existing groups. This is a between-subjects design. Recall the formula for the 95% confidence interval when a within-subject design was used:

$$CI_{95\%} = \overline{x1i - x2i} \pm t_{0.05} \frac{sd}{\sqrt{n}}$$

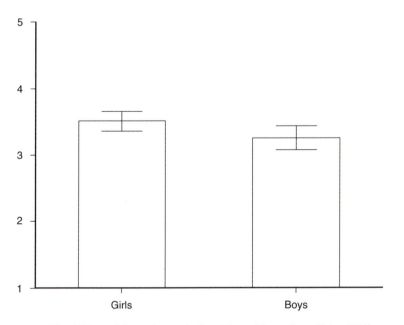

Figure 5.2 **The 95% confidence intervals for girls and boys (from Tatar 1998).**

where $x1i$ and $x2i$ are the two variables, the bar above denoting that this is the mean of this difference, and sd is the standard deviation of the difference. The formula for the between-subjects situation is conceptually similar. It is the difference between the two means $\pm\, t_{0.05}$ multiplied by the standard deviation divided by a function of the number of people. The tricky part is calculating the appropriate standard deviation and the number of people. It will be easier to work with the variances. The variances for the groups, call them var1 and var2, are combined into what is called the *pooled variance*:

$$\text{pooled var} = \frac{(n1 - 1)\text{var}1 + (n2 - 1)\text{var}2}{(n1 - 1) + (n2 - 1)}$$

where n1 and n2 are the sample sizes. If girls are group 1 and boys are group 2, and remembering to square the standard deviations to turn them into variances, you get

$$\text{pooled var} = \frac{(166 - 1)0.98 + (129 - 1)1.10}{(166 - 1) + (129 - 1)} = \frac{161.7 + 140.8}{165 + 128} = 1.03$$

This is a weighted average of the two variances. If you get a pooled variance that is not somewhere between the variances of the two groups, then something has gone wrong.

The formula for the 95% confidence interval for the difference is

$$CI_{95\%} = \overline{x1} - \overline{x2} \pm t_{0.05} \sqrt{\text{pooled var} \left(\frac{1}{n1} + \frac{1}{n2} \right)}$$

The new degrees of freedom are $n1 + n2 - 2$, or $166 + 129 - 2 = 293$. The t table does not go up this high. For values above 100 it is best to use $t_{0.05} = 1.98$ and $t_{0.01} = 2.63$. For Tatar's (1998) data, this is

$$CI_{95\%} = 3.51 - 3.26 \pm 1.98 \sqrt{1.03 \left(\frac{1}{166} + \frac{1}{129} \right)}$$

$$= 0.25 \pm 1.98 \sqrt{0.014} = 0.25 \pm 0.24$$

The 95% confidence interval goes from 0.01 to 0.49. Since all of the interval is positive, we can be confident that the mean for girls, in the population, is higher than for boys. If a 99% confidence interval was used the confidence interval is

$$CI_{99\%} = 3.51 - 3.26 \pm 2.63 \sqrt{1.03 \left(\frac{1}{166} + \frac{1}{129} \right)}$$

$$= 0.25 \pm 2.63 \sqrt{0.014} = 0.25 \pm 0.31$$

or from −0.06 to 0.56. The interval now includes negative values. Therefore, at this level of confidence we cannot say that the mean for the girls in the population is higher than the mean for the boys.

Summary

At the core of any scientific endeavour is inference, the ability to estimate population characteristics from observations of only a small sample of this population. This can be done poorly or well, depending on how well the sampling and allocation are done (see Chapter 4). When giving an estimate it is important not just to give a single point, but to convey information about the precision of the estimate. This is what confidence intervals are for. They allow you to say that you are confident, to a certain degree, that the population mean lies somewhere within the interval.

Ten years ago confidence intervals were often not reported. Usually there would be enough information (i.e. the mean, standard deviation and sample size) so that the reader could calculate the confidence interval. During the past decade many journal editors have stressed that confidence intervals should *always* be reported. This has rapidly affected practice and will continue to do so. Further, the American Psychological Association produced a Task Force Report (Wilkinson et al., 1999) that also argued that confidence intervals should be reported.

After reading this chapter you should have a conceptual grasp of why confidence intervals are used and be able to explain to other people what they mean when reported in newspapers. Further, you should be able to calculate the confidence interval for a mean, for the difference in means between two variables, and for the difference in means between two groups. The last of these is a little more complicated because – well, because the equations are longer. It is worth getting to grips with the equations in this chapter because they are the basis for the equations in the next chapter.

Exercises

5.1 Look in a newspaper to find *either* where a confidence interval is used and describe what it means, *or* where an estimate is produced without a confidence interval and describe what additional information the confidence interval would have provided.

5.2 For the coffee preference example, calculate the 99% confidence interval for the difference between the means. What conclusions do you make? Are the conclusions different from those made with the 95% confidence interval, and if so why?

5.3 Thirty-six female undergraduates were given a 15 minute maths test in groups of three (Inzlicht & Ben-Zeev, 2000). Half were in a group with two other females. Their mean was 70% correct. Half took the test in a room with two male confederates. Their mean was 55% correct. For both the standard deviation was about 20%. Calculate the 95% confidence intervals for these two groups, and for the difference between the two. What would you conclude from this study?

5.4 Cuc and Hirst (2001) asked Conservatives and Liberals in Romania if they could remember the prices of several common items (e.g. bread, butter, a national newspaper) from 1975 and 1985. The 1970s were a fairly prosperous period in Romania. This contrasts with the 1980s where the economic

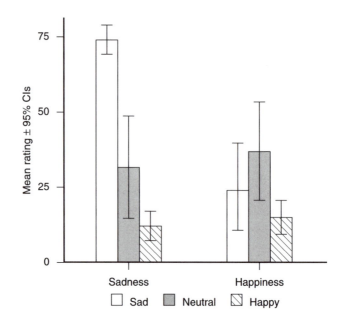

Figure 5.3   **The mean and 95% confidence intervals for ratings of sadness and happiness.**

Table 5.2   **Price estimation of common goods in Romania (Cuc & Hirst, 2001).**

	1975		1985		
	Mean	**sd**	**Mean**	**sd**	**n**
Liberal	0.00	0.26	0.27	0.62	49
Conservative	− 0.17	0.22	0.10	0.24	48

situation was dire. Price estimates from 48 Conservatives and 49 Liberals were compared with the true prices. They gave each person a price estimate score where 0 means accurate, positive scores mean overestimating the price, and negative scores mean underestimating the price. These, with the associated standard deviations, are given in Table 5.2. Calculate the 95% confidence intervals for each of these and display them in a graph.

5.5   Soon after Einstein proposed general relativity, nature presented scientists with an excellent opportunity to test it. There are three main hypotheses about how light will travel when near massive objects. First, it could be that light will not be affected, that its direction will not bend. The second is based on special relativity and the prediction is that it will bend 0.87″ as it passes the edge of the Sun. The third is based on general relativity and predicts the bend is 1.75″. Dyson et al. (1920; see Eddington, 1920, for a less technical account and one available in many university libraries) measured the bending of the light in two separate locations. The estimates for the bending of the light at the edge of the Sun were 1.98 ± 0.24″ in Brazil and 1.61 ± 0.60″ in West Africa.[2] What conclusions should the researchers have drawn?

5.6 Maccallum and colleagues (2000) took 24 highly hypnotisable participants, hypnotised them, and then told them to be sad, neutral or happy. They also had them go through a sad, a neutral or a happy story. Participants were then asked how happy they were and how sad they were (both on 0 (not at all) to 100 (extremely) scales). Figure 5.3 shows the means and 95% confidence intervals for these data. Discuss what this graph shows. Does it show that the manipulation works?

Further Reading

Wilkinson, L. and the Task Force on Statistical Inference, APA Board of Scientific Affairs (1999). Statistical methods in psychology journals: Guidelines and explanations. *American Psychologist, 54*, 594–604.
    This is an excellent report which stresses the value of confidence intervals, as well as other techniques.

---

2 These are not 95% confidence intervals. They are based on 'probable accidental errors'. As with standard errors, the conventional 'margin of safety [is] about twice the probable error on either side' (Eddington, 1920: 118). Therefore, for present purposes these can be treated as 95% confidence intervals.

The previous chapter described how to compute confidence intervals for means and differences between means. In some situations there are values that are of specific interest. In these cases scientists might wish to test whether the observed data are likely to have arisen if the population mean (or the difference between means) was this value. Usually it is if the difference between two means is zero. In this chapter you are taught two *t* tests, devised by 'Student,' who you were introduced to in Box 4.2. The first of these is that there is no difference between the means of two variables. It is called the *paired t test*. The second is that there is no difference in the mean of a variable for two groups. This is called the *group t test*.

The Paired *t* Test: Introducing Null Hypothesis Significance Testing

In the coffee preference example from Chapter 5 there is a special interest in the value zero for the difference. If the mean in the population were zero this would mean that overall both coffees are preferred equally. Some people may prefer the fresh coffee more, but others will prefer instant coffee. One way to explore the coffee data is to assume that the population mean is zero and to test how likely it would be to observe data as extreme as observed if the population mean is zero. This general approach is called *null hypothesis significance testing* (NHST) and the particular test used in this situation is a *paired t test* (also called related *t* test, matched t test and within-subject t test). Significance testing is a very common approach in the social and behavioural sciences, although it is controversial (see Box 6.1, and Harlow et al., 1997, for further details and http://www.indiana.edu/~stigtsts/ for a selection of quotes for and against significance testing). I will go through the coffee data to show how this is done.

If the two coffees were equally liked in the population and 10 people were asked to rate the two coffees, sometimes the mean difference rating in the sample would be above zero (fresh coffee better) and sometimes below zero (instant coffee better). Sometimes the mean would be *a lot* higher or lower than zero, but providing simple random sampling is used this is fairly rare. Significance testing quantifies how rare this is.

Significance testing is closely related to confidence interval construction. It involves finding the t value that would make the confidence interval just touch zero. Here is the equation for the observed t value:

$$t = \frac{\overline{DIFF_i}}{sd / \sqrt{n}} = \frac{1.0}{1.25/\sqrt{10}} = 2.53$$

It is the mean of differences divided by what is called the *standard error*. The standard error for a mean is the standard deviation divided by the square root of n. Standard errors are related to the precision of the estimate. The confidence interval was the mean plus and minus approximately two times the standard error. The t value is the mean difference measured in units of standard errors.

The next step is to find the probability associated with this t value. It is necessary to make some assumptions. The first assumption is that the data are independent, meaning the answers for one person are not dependent on the answers for another person. In the coffee drinking example, this would mean you should not test people in groups where one of them could spontaneously break into a 'its rich dark aroma' food critic mode, influencing other people's judgements. While there are methods to analyse clustered data (Goldstein, 1995; Wright, 1998a), they are beyond the scope of this book. The second assumption is that the distribution of the differences is normally distributed. You were introduced to the normal distribution in Chapter 2. Usually if the distribution looks roughly 'bell shaped' then this assumption is accepted. If this assumption is not met, then you should use the procedures described in Chapter 9.

Suppose that the mean difference in coffee preference is zero in the population (and the data are independent and normally distributed). If we take a sample of 10 people, we can find the t value. Sometimes the t value will be very small, other times it will be very large. Because random sampling is used, sometimes you will get a sample of primarily instant coffee lovers or fresh coffee lovers, even though

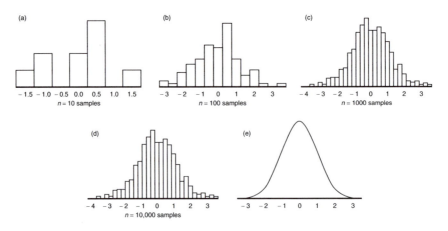

Figure 6.1(a–e) **Samples of size 10. As the number of samples increases the distribution gets more and more similar to the mathematical distribution shown in (e).**

the population mean difference is zero. It is because of random sampling that it is possible to quantify how likely this is.

Figure 6.1 shows a series of graphs. I created a population where the mean difference was exactly zero and that was normally distributed and took samples of size 10 from these. In Figure 6.1(a), I have taken 10 samples of 10 people, and graphed their *t* values in a histogram. As can be seen, some are above zero, some below. If there was no difference in population, and you ran 10 different studies, these are the sort of *t* values you might expect. The next graph (Figure 6.1b) shows what happened when I took 25 samples of size 10. As can be seen a general pattern is emerging. There are some fairly high scores, some low scores, and most near zero. As we move through 100, 1000 and 10,000 samples, it is clear we are getting something that looks like a bell-shaped curve. It is not quite the normal distribution, but is very close. It is called the *t* distribution with nine degrees of freedom. The mathematical form of it is in Figure 6.1(e). It is this *t* distribution listed in Appendix C and it is from this that the associated probability, or *p* value, is found. The *p* value is the probability of observing data as extreme as observed (or more), assuming the null hypothesis is true.

As was true with confidence intervals, where 95% is used as the convention, there is a convention for the critical *p* value. 'Critical' means that values have to be smaller than this to be felt so unlikely (assuming the null hypothesis is true) that the researcher would want to question whether the null hypothesis is true. If the *p* value is less than 5%, or 0.05, people tend to say it is 'significant'. This is an unfortunate choice of word because it does not have the same meaning as its English definition, but sadly it has stuck. It does not mean that the size of the

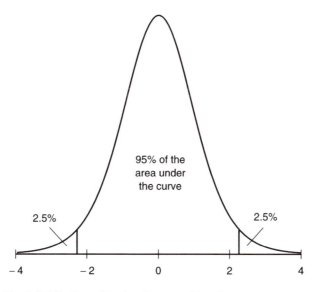

Figure 6.2   **The *t* distribution with nine degrees of freedom.**

difference is of any practical significance, only that if there were no difference in the population, then getting a result like the one observed is unlikely. A lot of people misinterpret the *p* value. A better word than 'significant' would be to say an effect was 'detected', if the *p* value was less than 5%, but this still does not perfectly convey the concept. Box 6.1 describes what *p* is and what it is not in more detail.

In Figure 6.2 the area under the curve, on each side, that includes 2.5% of the total area under the curve, has been shaded. This corresponds approximately to $t = -2.26$ and $t = 2.26$ for 9 degrees of freedom. If the value is either less than -2.26 or greater than 2.26 the null hypothesis is rejected at the 5% level.

An obvious question is: how well does this work? Well, in the group of 10,000 samples, 257 had a *t* value below -2.26 and 232 had a *t* value above 2.26. Thus, for 4.89% of the samples we rejected the null hypothesis (even though it was true). This is known as a Type I error, falsely rejecting the null hypothesis. While it is best to avoid errors, this result is actually about what we expect. By using a 5% level, we would expect that when the null hypothesis is true, it will be falsely rejected about 5% of the time (and 4.89% is close to 5%). If 5% was too high (i.e. if there were grave consequences of falsely rejecting the null hypothesis), then a more stringent level could be set, like a critical *p* value of 1% or 0.01. For the *t* distribution with nine degrees of freedom, the critical points are below -3.25 and above 3.25. In total, there are 101 of the 10,000 samples (1.01%) outside these points. Using the 1% level lowers the chance of a Type I error, but as we will see later it increases the chance of another error, creatively called a Type II error. This is where there is an effect, but the p value is a non-significant, and we fail to detect the effect.

Taking lots of samples as is done in Figure 6.1 demonstrates the concept of a frequentist probability: that in the long run we should make about the right number of Type I errors. However, in most circumstances there is just a single trial, and the t table is used.[1] Appendix C shows what the observed t values must exceed in order to reject the null hypothesis at the p values of 5% and 1%. If the observed value either exceeds the critical value or does so when ignoring the minus sign then you know that the associated probability value, referred to as p value, is lower than 5% and/or 1%. Here we see that 2.53 is larger than 2.26 (the degrees of freedom are still nine), so that the probability level is less than 5%. However, the observed t is smaller than 3.25, so the probability is not smaller than 1%. You would report $p < 0.05$. Appendix C does not allow you to be more precise.

To get a more precise p value then you have to use a computer (or a more detailed table). Most statistical packages will compute precise p values from an observed t and the degrees of freedom. The more precise p value for this example is 0.032. This is what you should report when available as it provides more information than simply saying $p < 0.05$ (see Wright, 1999, for more details).

BOX 6.1   WHAT IS p?

Almost every social and behavioural science journal is filled with p values. Significance testing is taught in most courses as if it is uncontroversial, straightforward and conceptually simple. However, p is a difficult concept. People are often taught a watered down incorrect meaning of p and this leads to much misinterpretation and in the end increases confusion. After taking statistics courses students (and some staff) think the p value is the probability that the null hypothesis (that there is no difference in the population) is true. Another popular definition is the opposite: that p is the probability that the null hypothesis is false. Neither of these is the correct definition. In fact the null hypothesis is always false. In the entire population the chance of the mean difference in coffee preference, for example, being *exactly* zero is zero. To quote one of my favourite statisticians (yes, people like me have favourite statisticians), 'the null hypothesis … is always false in the real world' (Cohen, 1990: 1308).

---

1 Actually, there are modern statistical techniques, for example *bootstrapping*, which use a similar technique involving taking many samples from a distribution, and seeing how often extreme scores arise. These are more advanced and beyond the scope of this book (Efron & Gong, 1983).

p is *not* a probability about any hypothesis. It is the probability about the data, assuming that the null hypothesis *is* true. In the examples in this chapter you first draw (or at least think about) the *t* distribution that would arise if the null hypothesis was true. You are assuming it is true. Then you look for the probability of observing a mean of data as extreme as observed. I know I am repeating myself, but I want this message to be clear. The *p* value is the probability of data, as extreme as observed (or more extreme), assuming that the null hypothesis is true.

I know this is a difficult concept. Most people who think that they understand the concept of *p* will confidently give an incorrect answer when asked what *p* is the probability of. If people knew (and understood) the correct meaning of *p* they would probably use confidence intervals more and *p* values less. They would probably also use more graphs and be more careful to report means, standard deviations and other descriptive statistics. Among methodologists, there is agreement that *p* values are overused. This is nicely summarised by one of the more vocal and well-respected critics: 'the almost universal reliance on merely refuting the null hypothesis is a terrible mistake, is basically unsound, poor scientific strategy, and one of the worst things that ever happened in the history of psychology' (Meehl, 1978: 817). Yet, it is the dominant approach.

The use of *p* values is hotly debated. A recent statement by the American Psychological Association's Task Force on Statistical Inference (Wilkinson et al., 1999) makes clear that there are lots of ways to look at data. I think good statistical practice is best summarised by a statement made in an interview with one of the co-chairs of that Task Force, Robert Rosenthal, that researchers should 'make friends with their data'.

Let's consider another example from Chapter 5. The main purpose of Newton's research was to examine differences in hostility between when inmates entered Grendon Prison and when they were discharged. At discharge the HDHQ was also administered (the mean was 21.6 with a standard deviation of 9.2), and Newton compared these with the earlier scores. She subtracted the scores for each person and found the mean of this difference was − 6.6 with a standard deviation of 9.0. To do a *t* test you simply plug the numbers into the *t*-test equation:

$$t = \frac{\overline{DIFF_i}}{sd / \sqrt{n}} = \frac{-6.6}{9.0 / \sqrt{94}} = -7.11$$

With a *t* value of − 7.11, and 93 degrees of freedom, it surpasses the critical value for *p* of 5% and 1%. Its actual *p* value is very close to zero, though still slightly above zero. Statistical packages often output $p = 0.000$. Instead of writing this, it is better to write $p < 0.001$ so that the reader knows that you know it is still above zero. Thus, observing a *t* value this large in magnitude is extremely unlikely if the null hypothesis was true, but theoretically it is still possible. However, because it is so unlikely, we reject the null hypothesis that the population difference is zero, and say that hostility decreased.

Comparing Two Groups (Assuming Equal Variances)

Recall the study by Tatar (1998), introduced in Chapter 5. In the sample of Israeli secondary school students there were 166 girls and 129 boys, who were asked to rate (on a 1 to 5 scale with 5 being high) how much 'makes me learn willingly' characterises a 'significant' teacher. The mean for girls was 3.51 with a standard deviation of 0.99. The mean for boys was 3.26 with a standard deviation of 1.05. The first thing to do is to graph the confidence intervals for both of these groups using the procedures described in Chapter 5.

The question usually asked is: 'are girls' and boys' attitudes the same or does one gender feel that "makes me learn willingly" is a more important attribute for a "significant" teacher?' The way that this is usually evaluated is with a group t test (sometimes called a between-subjects t test). Conceptually, it has a very similar format to the paired t test and computationally it has similarities to constructing a confidence interval for group differences in Chapter 5. It is the difference between the means of the two groups divided by the standard error of this difference. However, the way that the standard error is calculated is different from the paired t test. There are a few ways to do a group t test depending on which assumptions you are willing to make.

First, I will assume that the standard deviations (or the variances) are the same in the populations from which the two samples are drawn (i.e. Israeli boys and girls). It is easier to work with variances because it makes the calculations slightly simpler (recall that squaring the standard deviation results in the variance). As with Chapter 5 let var1 and var2 be the variances of the two groups and n1 and n2 their respective sample sizes. Then as with Chapter 5 the 'pooled variance' is

$$\text{pooled var} = \frac{(n1 - 1)\text{var1} + (n2 - 1)\text{var2}}{n1 + n2 - 2}$$

This is a weighted mean of the two sample variances. If one sample is larger, it is given more weight. If the two sample sizes are equal (i.e. n1 = n2), then it is simply the mean of the two variances. This is then put into an equation for t:

$$t = \frac{\overline{x1} - \overline{x2}}{\sqrt{\text{pooled var} \, (1/n1 + 1/n2)}}$$

From Chapter 5 we know that pooled variance is 1.03 for Tatar's data. Putting this into the t test equation yields

$$t = \frac{3.51 - 3.26}{\sqrt{1.03(1/166 + 1/129)}} = \frac{0.25}{\sqrt{0.014}} = \frac{0.25}{0.12} = 2.08$$

The next step is to see whether this is statistically significant. The t table is used (Appendix C). The t table requires knowing the degrees of freedom (df). The degrees of freedom for a group t test, when equal variances is assumed, are

$$df = n1 + n2 - 2$$

So for this example $df = 293$. Being conservative we can use the $df = 100$ row and we see that the value exceeds 1.98 and therefore we can say that the difference is significant at a critical p of 5%. However, it does not exceed the value for 1%. While statistically significant (at 5%) it is important to think about the size of the effect. A difference of 0.25 on a five-point attitude scale is not large regardless of any statistical significance. With large samples trivial effects may become statistically significant. It is important to look at the confidence intervals, as discussed in Chapter 5.

The equations shown in this section are suitable when the standard deviations, or variances, of the two groups are approximately the same.

Not Assuming Equal Variances: Acupuncture

There is much debate about the therapeutic value of acupuncture. Numerous theories have been put forward for why acupuncture works. As scientists we need data to help to judge the therapeutic value. Given the popularity of acupuncture, there has been surprisingly little empirical research using proper scientific methods to assess its value.[2] One exception is by Allen et al. (1998). Their sample was composed of depressed women who were either randomly assigned

---

2 For a powerful book describing the lack of research demonstrating how different therapies are better or worse than control conditions, see Dawes (1994).

to an acupuncture condition ($n1 = 12$) or put on a waiting list ($n2 = 11$). This is a common 'control' group in such studies, though it raises some ethical issues.[3]

Participants were given a scale for measuring depression both when they arrived and after 8 weeks. The authors calculated a change score where negative scores mean that the person became less depressed (i.e. negative scores are good). The mean and standard deviation for the acupuncture group were − 11.7 and 7.3. The equivalent scores for the waiting list group were − 6.1 and 10.9. While the standard deviations in the students' attitudes example were fairly close, these standard deviations are more different. A rough rule of thumb is that if the ratio of the larger variance to the smaller variance is greater than 2 then you should be concerned. Here, $10.9^2/7.3^2 = 118.8/53.3 = 2.23$. The procedures in the last section assumed that the variances were equal. Here I describe the procedures to use when this assumption is not made.

Again, it is best first to calculate the confidence intervals and to graph these. The confidence intervals are

$$\text{for acupuncture: } CI_{95\%} = -11.7 \pm 2.20\frac{7.3}{\sqrt{12}} = -11.7 \pm 4.6$$

$$\text{for waiting: } CI_{95\%} = -6.1 \pm 2.23\frac{10.9}{\sqrt{11}} = -6.1 \pm 7.3$$

The improvement was better (i.e. the depression scores dropped more) for the acupuncture sample compared with the waiting list sample, but the difference appears small. A group t test can be used to see if the difference is statistically significant. There are two popular approaches to running a t test when the variances are different, one to use when you are doing the calculations yourself and the other when a computer is doing the calculations. Here only the 'by hand' approach will be discussed. You calculate t as follows:

$$t = \frac{\overline{x1} - \overline{x2}}{\sqrt{\text{var1}/n1 + \text{var2}/n2}}$$

$$t = \frac{-11.7 - -6.1}{\sqrt{53.3/12 + 118.8/11}} = \frac{-5.6}{3.90} = -1.44$$

---

3 There was a third group given a non-specific treatment. Despite random allocation this group had a much lower initial mean than the other groups. Therefore this group will not be considered here.

Next, you have to calculate the degrees of freedom. The 'by hand' method is to take the smaller of $(n1-1)$ and $(n2-1)$. Because $n_2 < n_1$, we use $df = 11-1 = 10$. Next, go to the *t* table. In the 10th row it shows that the critical value p at 5% is $t = 2.23$. As the observed value does not exceed this value we fail to reject the hypothesis that the values of the treatments are identical. This is a conservative method, meaning that it is less likely that the value will be statistically significant than the computer approach. The approach used by many computer programs calculates the degrees of freedom in a more complex way. It is best to use those values if available. Do not worry that the number of degrees of freedom is not a whole number.

There were a lot of new equations in Chapter 5, so I did not include a 95% confidence interval for the difference between the means of two groups when the variances are not assumed to be equal. The equation is

$$CI_{95\%} = (\overline{x1} - \overline{x2}) \pm t_{0.05}\sqrt{\frac{var1}{n1} + \frac{var2}{n2}}$$

which here yields

$$CI_{95\%} = -5.6 \pm 2.23(3.9) = -5.6 \pm 8.70$$

which includes zero, the population mean assumed for the null hypothesis.

Summary

In this chapter you were introduced to null hypothesis significance testing and two statistical tests: the paired and the group *t* tests. The paired *t* test is used when you are comparing the mean of two variables. The group *t* test is used when you want to compare the means of two separate groups of people. It is important before you do either of these that you look carefully at your data (i.e. make friends with your data). It is worth graphing the variables' distributions (Chapter 2) and constructing confidence intervals (Chapter 5). Too often people leap into a statistical test simply wanting to see if it is statistically significant. As discussed in Box 6.1, finding a significant p value is not as important as most social scientists believe.

6.1 Fisher and Geiselman (1992) developed what is called the *cognitive interview*. It is a method for interviewing eyewitnesses that has been shown to elicit more accurate information compared with the standard police interview. In a hypothetical study, suppose 10 people were trained in the cognitive interview and 10 other people were trained in the standard police interview. Twenty participants were shown individually a video of a bank robbery. The first 10 participants were interviewed by the 10 cognitive interviewers. The numbers of accurate details recalled were

20, 22, 17, 32, 16, 9, 18, 19, 23, 24

The second 10 were interviewed by those trained in the standard police interview. The numbers of accurate details recalled were

17, 11, 21, 9, 14, 23, 10, 14, 20, 11

What is the evidence for there being a difference between the two interview techniques?

6.2 A political researcher was interested in UK and French people's attitudes towards the single European currency, the 'Euro'. Suppose that 20 people in each country were asked 'Do you think that the Euro will be good for your country?' using a 1 to 7 scale where 1 is anti-Euro and 7 is pro-Euro. The mean in the United Kingdom was 2.2 with a standard deviation of 3.0, while the French had a mean of 5.2 with a standard deviation of 2.0. Are means of these samples sufficiently different to reject the hypothesis that the UK and French people have the same attitude towards the Euro?

6.3 (a) Find a student from the social, behavioural or physical sciences who has done statistics (probably avoid maths and stats students) and who has used *p* values. Ask the student what s/he thinks *p* is the probability of. Write down what the student said and say whether s/he is right or wrong.

(b) Find a student who has taken no statistics. Explain to this student what a *p* value is and why you use it. Write down the problems you had explaining it to the student.

6.4 From the data in Table 5.1, test the hypothesis that instant and fresh coffee are equally well liked.

6.5 Recall the example from Tatar (1998) from Chapter 5. She compared boys (*n* = 129) and girls (*n* = 166) on whether they felt 'makes me learn willingly' was a characteristic of a 'significant' teacher (on a 1 to 5 scale with 5 being agree). The mean for girls was 3.51 with a standard deviation of 0.99. The mean for boys was 3.26 with a standard deviation of 1.05. Conduct a *t* test on these data and describe what you would conclude.

6.6 A drug company has created a drug that it feels can increase people's memory. The company randomly allocated a group of 20 people into two groups. One group of 10 people received a pill containing the drug while the other group received a pill without the drug (what is sometimes called a *placebo*). Participants were then given a memory test, scored from 0 to 10 where higher scores stand for having better memory scores. The data are:

Placebo group	2	5	4	4	1	3	4	10	4	4
Drug group	5	8	10	3	5	1	1	9	8	8

What are the means for the two groups? Does one group have a higher mean than the other?

6.7 Rothblum and Factor (2001) were interested in differences between lesbians and their heterosexual sisters on a number of characteristics. They had 173 lesbian–heterosexual pairs of sisters fill out the Rosenberg Self-Esteem Scale. In this study they compared using a within-subject design versus a between-subjects design. The mean for lesbians was 33.9 with a standard deviation of 4.55. The mean for the heterosexuals was 32.0 with a standard deviation of 5.04. Conduct a between-subjects *t* test using these means and standard deviations.

Rothblum and Factor realised that self-esteem might relate to both genetic and environmental factors. They felt that taking pairs of lesbian–heterosexual sisters and subtracting the scores for each pair (the pair is now the unit of analysis) would help to control for some of these additional factors.[4] If you subtract the heterosexual sister's score from the lesbian sister's score, you get a mean of 1.9 with a standard deviation of 7.00. Calculate the within-pair *t* test for this mean and standard deviation.

---

4 This is the same conclusion 'Student' (1931) comes to in his criticism of the Lanarkshire milk study (see Box 4.2). He said how it would have been much more effective to use siblings; he actually suggests monozygotic twins as participants. Here it appears that the genetic and sibling environmental effects were not large as the effects are of a similar size.

Further Reading

Cohen, J. (1990). Things I have learned (so far). *American Psychologist, 45,* 1304–1312.

    This is a great article by one of the most important psychology statisticians. It is informative and well written, elegantly presenting wisdom gained through years of research.

Cohen, J. (1994). The Earth is round ($p < .05$). *American Psychologist, 49,* 997–1003.

    This paper is critical of null hypothesis significance testing.

Wright, D. B. (1999). Science, statistics and three 'Psychologies'. In D. Dorling & L. Simpson (Eds.), *Statistics in Society: The Arithmetic of Politics* (pp. 62–70). London: Arnold.

    In this chapter I criticise how statistics are used in psychology, as well as how science, more generally, is used by psychologists.

7 Comparing More than Two Groups: ANOVA

In Chapter 6 you were shown how to compare the means of two variables and of two groups using null hypothesis significance testing. Sometimes you want to compare the means of more than two groups. While this can be done using a series of t tests, there is a special procedure, called analysis of variance or ANOVA, which has been designed for this purpose. The assumptions, like normality and having similar variances, are basically the same as with the between-subjects group t test. Here we will assume that the variances are equal. The ANOVA procedure is important also because it is easily extended to more complex designs. It is a very popular procedure that is reported in many journal articles. In this chapter you will learn how to do what is sometimes called a *one-way ANOVA*. This is the simplest type. If you go on in statistics more complex ANOVAs will be introduced.

An Example: Cognitive Dissonance

In one of the classic studies of social psychology, Festinger and Carlsmith (1959) had participants spend about an hour putting spools onto a tray and turning square pegs a quarter rotation. It was designed to be boring and succeeded. After participants finished this tedious task, the experimenter pretended as if the study was over and gave them a pretend debriefing. Participants were told that there were two groups in the study. Half of the people were only given the basic instructions. The participants were told that they were in this condition. The other half were told by a confederate (someone working for the experimenter but pretending to have just taken part in the study as a participant) that the experiment was enjoyable. At this point, the real study was just beginning.

There were three groups in Festinger and Carlsmith's study. There was a control group who after the 'debriefing' were ushered into a waiting room. There were two experimental groups. For each of these the experimenter explained that the usually reliable confederate had phoned saying that he could not make it. The experimenter asked if the participant would help out and tell a female participant who was waiting in the next room that the experiment was enjoyable. One group was paid $1 and the other was paid $20.[1] Most complied, although a few said they were suspicious and their data were discarded.[2] After the participant either waited in an empty room (control group) or told the confederate that the boring task was enjoyable, they thought they were done. On leaving the building, they were informed that the department monitors all experiments and asked if they would fill out a questionnaire. This in fact was an integral part of the study. It included a question asking them, on a − 5 to + 5 scale, how interesting and enjoyable the study (the boring spools and pegs tasks) was. The prediction from Festinger's cognitive dissonance theory is that those paid only $1 were more likely to say the task was enjoyable compared with the other groups.

Festinger and Carlsmith had 20 people in each condition. I have recreated these data so they closely resemble their original data. Table 7.1 gives the data and some calculations. The first step in analysing these data would be to look at the distributions, perhaps doing a boxplot and some histograms (see Chapter 2). After this you would probably want to find the 95% confidence intervals and plot these, as is shown in Figure 7.1. It shows that the highest level of enjoyment was for those given only $1 (enjoyment ratings were taken before they were asked to return the money).

ANOVA is used to examine differences among means. The null hypothesis is that the means are the same in the population for each of the three groups. Conceptually it works by dividing the total variation among all the data points into variation within each condition and variation between the conditions.

Total variation = within-condition variation + between-conditions variation

---

1 All were asked for this money back at the end of the real experiment. Festinger and Carlsmith (1959: 207) said all 'were quite willing' to do this. I do not know if this aspect would replicate in the present day.

2 Another participant's data were discarded because he asked for the female's phone number and said 'he would call her and explain things' (p. 207) and wanted to stay around until she was done with the experiment so they could talk, presumably wanting to 'debrief' the young lady (who was the actual confederate). After the actual experiment, participants were debriefed with the female confederate present. This participant must have felt pretty embarrassed.

Table 7.1 **Data recreated to match very closely Festinger and Carlsmith's (1959) classic study of cognitive dissonance.**

		Condition			
Control (mean = − 0.45)		$1 (mean = 1.35)		$20 (mean = − 0.05)	
Value	$(x_i - \bar{x})^2$	Value	$(x_i - \bar{x})^2$	Value	$(x_i - \bar{x})^2$
0	0.20	3	2.72	1	1.10
− 3	6.50	1	0.12	2	4.20
3	11.90	1	0.12	3	9.30
2	6.00	3	2.72	0	0.00
− 2	2.40	2	0.42	1	1.10
− 1	0.30	3	2.72	3	9.30
2	6.00	3	2.72	0	0.00
3	11.90	2	0.42	− 2	3.80
− 3	6.50	2	0.42	2	4.20
− 5	20.70	2	0.42	1	1.10
2	6.00	2	0.42	0	0.00
− 3	6.50	2	0.42	0	0.00
3	11.90	− 4	28.62	− 1	0.90
0	0.20	4	7.02	− 2	3.80
− 2	2.40	0	1.82	− 1	0.90
− 2	2.40	− 3	18.92	− 4	15.60
− 2	2.40	4	7.02	− 3	8.70
− 2	2.40	1	0.12	− 1	0.90
− 1	0.30	1	0.12	0	0.00
2	6.00	− 2	11.22	0	0.00
	$\Sigma = 112.95$		$\Sigma = 88.55$		$\Sigma = 64.95$
	var = 5.94		var = 4.66		var = 3.42
	df = 19		df = 19		df = 19

The 'total variation' is found by subtracting the overall mean (0.28 for the data in Table 7.1) from each score, squaring this difference, and summing all this. It is often called the *total sum of squares* and is 302.18. If you divided by $n - 1$ (59 here) you get the variance. The number 59 is also the degrees of freedom of the total variation.

The next step is calculating the within-condition sum of squares. This is done in Table 7.1. The columns marked 'Value' give the value for each person in each condition. $(x_i - \bar{x})^2$ gives the squared deviations from the mean within the condition. In the bottom row the sums of these deviations are reported. Adding these three values together produces the within-condition variation:

$$\text{Within-condition variation} = 112.95 + 88.55 + 64.95 = 266.45$$

There is also the number of degrees of freedom associated with this sum of squares; it is $n - k$, where n is the total sample size (60) and k is the number of conditions (3).

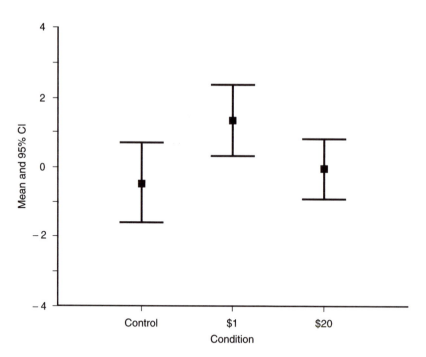

Figure 7.1 **The means and 95% confidence intervals for Festinger and Carlsmith's (1959) classic demonstration of cognition dissonance.**

Next, the between-conditions variation is calculated.[3] Because the total variation comprises the within-condition variation and the between-conditions variation,

Between-condition variation = total variation − within-condition variation

$$35.73 = 302.18 − 266.45$$

The between-conditions variation is that attributable to variation, or differences, among the condition means. The larger this difference the more variation among means. It is useful to calculate the percentage of total variation that is within-condition variation and the total that is between-conditions variation. Here, the within-condition variation is 266.45/302.18, which is 88% of the variation. That leaves 12% of the total variation. Thus, we can say that 12% of the total variation is attributable to differences among the means of the conditions. I will refer to this as R^2, although it is also often referred to as η^2 (eta squared).

3 The between-conditions variation can also be calculated directly. For each condition subtract the overall mean from the condition mean, square the difference, multiply this by the number of people in the condition and sum up these three values. I have relegated this to a footnote because rounding errors tend to be exaggerated.

People often say '12% of the variation is explained' by the difference among conditions. I try to avoid the word 'explained' because no statistical test 'explains' a finding – that is up to the scientist. It is important to realise what statistics can and cannot do, and people should be careful in the way they describe their results.

You could stop here. Finding that 12% of the variation is accounted for by between-conditions variation is important. However, often researchers want to know if there is a statistically significant difference among the conditions. As with the t test, the first step is to assume that in the population the means of each condition are the same. When people are allocated to the different conditions, by chance there is likely some differences among the means. If the means in the population are the same, then hypothesis testing allows us to say how unlikely having large differences is (assuming the means are all the same in the population).

Dividing the within-condition variation by the associated degrees of freedom provides an estimate for the variance *if* there are no group differences. The degrees of freedom for this variation are the total number of cases, minus the number of groups. Here that is $60 - 3$ or 57. This estimate is called MSe, which stands for mean sums of squares error. By error it means variation within the conditions (so I could have labelled it MSw, but the convention is MSe).

$$\text{MSe} = \frac{\text{within-condition variation}}{df_{\text{within}}} = \frac{266.45}{57} = 4.67$$

Similarly, if the means in the population are equal, then another estimate for the variance is the mean sums of squares between conditions (MSb). Here the degrees of freedom are the number of conditions minus one ($3 - 1 = 2$). The within-condition df plus the between-conditions df equal the total df: $(n - k) + (k - 1) = (n - 1)$.

$$\text{MSb} = \frac{\text{between-conditions variation}}{df_{\text{between}}} = \frac{35.73}{2} = 17.87$$

If the null hypothesis, that the means in the population are the same, is true, then these two should be estimating the same thing. If they are estimating the same thing, then the ratio of one to the other should be about one. If the differences among the groups are larger than would be expected, then MSb will be larger than

MSe. Therefore, we divide MSb by MSe. This is called the value F, after the famous statistician Sir Ronald Fisher:

$$F(2, 57) = \frac{MSb}{MSe} = \frac{17.87}{4.67} = 3.83$$

If this is greater than one, as here, it means that the means differ more than would be expected by chance, assuming the population means are the same. If $F < 1$, the means differ less than expected. The question is how large a difference is necessary in order to reject the null hypothesis. As with the t tests, the F value can be looked up in an F table. One such table appears in Appendix D. It is necessary to know the degrees of freedom for the numerator (the number on the top) and the denominator (the number on the bottom). This table gives the critical values for a critical p of 5% and 1%. If we go to the second column and then go down to the row for $df = 50$, the two critical values are 3.18 and 5.06. Therefore, we can say that the means are different at the 5% but not at 1% level.

With growing technology, most statistical procedures are now run with the aid of a computer. Most programs produce what is called an ANOVA table. The following comes from SPSS:

	Sum of Squares	df	Mean Square	F	Sig.
Between Groups	35.733	2	17.867	3.822	.028
Within Groups	266.450	57	4.675		
Total	302.183	59			

As with the t test, because the computer gives you a precise p value it is best to report that. So here you would say $F(2,57) = 3.82$, $p = 0.03$ and that 12% of the total variation can be attributed to differences among the conditions.

At this point you can conclude that there is a difference among the means, but strictly speaking you should not say which means differ from which. To do this you have to do what is called *multiple comparisons*. There are entire books devoted to this subject (e.g. Toothaker, 1993). Most of them involve comparing each of the conditions with each other, but requiring a lower p value to say that the difference is significant. The number of conditions determines how much lower the p value needs to be. The more conditions, the lower the p value. Here there are three conditions. There are three possible pairwise comparisons.[4] Rather than get

---

4 These are control v. $1, control v. $20 and $1 v. $20. When there are four conditions, there are six comparisons. When there are five conditions, there are 10 comparisons. When there are 10 conditions, there are 45 comparisons. Thus, the number of comparisons rapidly increases. In general, dividing 5% by the number of comparisons is a conservative approach, meaning that you may fail to detect many differences.

bogged down in details, I will simply say that you should lower your critical *p* value somewhat, depending on how many comparisons there are. One approach is to lower it by dividing by the number of comparisons (0.05/3 = 0.017).

If *t* tests (which is what Festinger and Carlsmith, 1959 did) are done, the difference between the control and the $1 groups has $t(38) = 2.47$, $p = 0.02$, between the control and the $20 groups has $t(38) - 0.59, p - 0.56$, and between the $1 and $20 groups has $t(38) = 2.20, p = 0.03$. If we set a critical level of 0.017 then none of these would be statistically significant. As discussed in Chapter 6, most methodologists believe that *p* values are not the most important thing in deciding whether to accept or to reject a hypothesis. Here, the direction and size of the effect and the compelling predictions by Festinger and Carlsmith mean that most people would accept that the $1 group is higher. Looking at a graph with confidence intervals is much more important than relying on *p* values (see Figure 7.1).

Thatcher's Resignation

A common exercise for Anova questions is being given part of an ANOVA table and having to fill in the blanks. A few years ago I worked on some projects looking at people's memories and feelings about Margaret Thatcher's resignation as Prime Minister of the United Kingdom (Wright et al., 1998). One of the questions we asked was how important people thought the event was. We looked to see how the importance ratings, which were done on a 0 to 4 scale, where 0 was 'not important at all' and 4 was 'extremely important', differed for various social groups. We looked at lots of things, but one thing we did not look at was whether importance ratings differed by marital status. Figure 7.2 shows the 95% confidence intervals for four different groups: married (or co-habiting), single, widowed and divorced (or separated).

As can be seen, the mean for married is the highest, and the mean for widowed is the lowest. Below is the ANOVA table for these data. I have removed most of the numbers, and replaced them with letters. Try solving these for yourself, before reading further.

	Sum of Squares	df	Mean Square	F	Sig.
Between Groups	A	B	D	F	.001
Within Groups	2565.771	2106	E		
Total	2586.162	C			

From using this example with other students, many of you might have had problems with the degrees of freedom (the 'df' column). If you had the value for B, then you could calculate C by adding 2106 to B. Similarly, if you had C, you

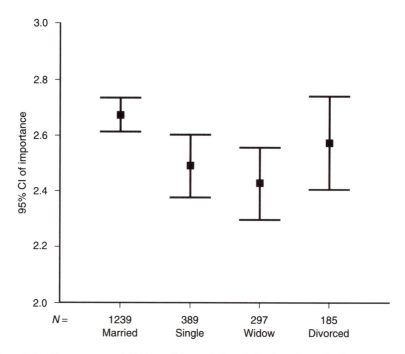

Figure 7.2   **The means and 95% confidence intervals for how important an event Thatcher's resignation was, broken down by marital status.**

could calculate B by subtracting 2106 from it. The answer is that the *df* for the between-groups sum of squares is equal to the number of conditions minus one, so $4 - 1 = 3$. So, B = 3, and therefore the value for C is $3 + 2106 = 2109$.

A, the between-groups variation (or sums of squares), can be calculated by subtracting the variation within the groups (2565.771) from the total variation (2586.162), to produce 20.391. The values for D and E are the respective sums of squares divided by their degrees of freedom, so $20.391/3 = 6.797$ and $2565.771/2106 = 1.218$. The *F* value is the mean square for the between-groups divided by the mean square for the within-groups, $6.797/1.218 = 5.580$. This is significant. It is important to note that the sum of squares values, the degrees of freedom, the mean square values, the *F* value and the significance level all have to be positive. If you get a negative number you have made a mistake.

Because the *F* value is significant, the next step is usually to see which groups differ. Figure 7.2 shows that the mean for the married (and co-habiting) group is the highest, and its confidence interval does not overlap with the single or widowed group. *t* tests were run on each of the six possible pairs. Table 7.2 gives the *t* values and their associated *p* values. Because six tests are done, this increases the chance that a Type I error (falsely rejecting a hypothesis)

Table 7.2 **Pairwise *t* tests and their associated *p* values comparing self-rated importance of Margaret Thatcher's resignation as Prime Minister of the United Kingdom. Because there is more than one test, a more stringent critical level should be used. Here, I suggest 0.01 would be good.**

	Marital status		
	Single	Widowed	Divorced (separated)
Married (co-habiting)	2.90 (0.004)	3.48 (0.001)	1.16 (0.245)
Single		0.70 (0.484)	0.83 (0.405)
Widowed			1.37 (0.171)

occurs. Therefore we would probably only be convinced that the difference exists if p is less than about 0.01.[5] However, this increases the chance of missing a real effect, a Type II error. There are two comparisons which are statistically significant, the difference between married and single, and the difference between married and widowed.

Now at this point the researcher would be tempted to say that these effects are large – after all the p values are 0.004 and 0.001. However, the p values do not tell you how large the effects are. They just say whether an effect has been detected. If you want to say how big an effect is you have to look at the actual difference. The y axis in Figure 7.2 only shows a very small area of the total. Figure 7.3 shows the 95% confidence intervals showing the entire range of possible scores, from 0 to 4. This still shows that the married responses are higher, but the differences look smaller. After all, the highest mean (2.67 for married) is less than a quarter of a unit on the 0 to 4 scale above the lowest mean (2.43 for widowed).

The confidence intervals tell you how precise your estimate of the mean is. When you increase the sample size confidence intervals tend to get smaller. It is important to estimate the size of the effect, to say what proportion of the overall variation is accounted for by the difference between the groups. If we divided 20.39 by 2586.16 we would get $R^2 = 0.8\%$. Less than 1% of the variation can be attributed to differences in marital status. It is important to stress that finding something that is statistically significant does *not* mean that it is either a big or an interesting effect. This will be expanded upon in Chapter 8.

---

5 I got 0.01 because 0.05/6 is 0.008, which is about 0.01. An obvious question is why don't I use 0.008? I could have. The reason that I did not was because I wanted to stress that there is no single correct value. Using 0.008 is not better than 0.01 or even 0.02. Using 0.02 would increase the chances of Type I errors, but decrease the chances of Type II errors.

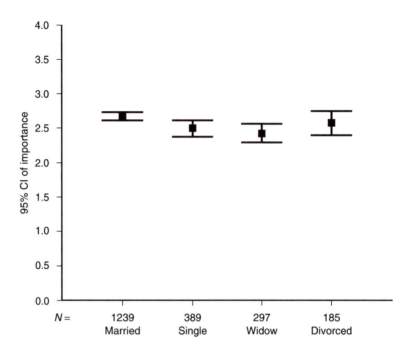

Figure 7.3 **The means and 95% confidence intervals for how important Thatcher's resignation was for people of different marital status. The y axis now shows the entire range of responses.**

Summary

I have a confession. It took me a long time to understand ANOVA. I had a really good teacher for this, so it was something to do with me. Perhaps it was because there was all this talk of variation and variance when what I wanted to know was if there was a difference among the means. It took a while for me to get it. The difference among means is measured by their variation. In a between-subjects t test, you look at the difference between two means. If the two means differ by a lot, then that set of two means has a large variation.[6] If you have three or more means, then the more different they are from each other, the larger the variance for them. In both the t test and ANOVA you then divide this by a measure of how precise the estimate is. In both cases it is based on the amount of within-group variation and the sample size. As the within-group variation increases, the estimate becomes less precise (and t and F get smaller, and the confidence

6 A t test is just a special case of an Anova. If you run both a t test and an ANOVA on the same set of data you will get the same p value and t^2 will be equal to F.

intervals get larger). As the sample size increases, the estimate becomes more precise (and t and F get larger, and the confidence intervals get smaller). I guess the main point of this confession is that ANOVA is about variance, but variance measures the differences among the means. Do not worry too much if any particular statistical procedure does not seem self-evident at first.

The one-way Anova, described in this chapter, can be extended to more complicated designs (see Wright, 1997b, for a little more extension, and Stevens, 1996, for a lot more; also see Field, 2000, for conducting ANOVAs on SPSS). ANOVA involves more calculations than the other techniques previously discussed. It is important to get them right. If you are doing ANOVA with a computer, you will get lots of output. Some of it will be of an ANOVA table. It is worth making sure you understand what all the terms mean.

Once you find a statistically significant F value, you then want to know which groups differed from which. Looking at a graph of the confidence intervals will be very valuable, but you will probably also want to run some additional statistical tests. The number of comparisons increases very rapidly as the number of groups increases. There are certain rules when you have multiple comparisons. The main rule is, because these give more chances to have a Type I error, the critical p value should be lowered. There are dozens of tests for multiple comparisons that all, in certain ways, do this. Most of the main computer statistical packages will offer several options. Some textbooks will go on for pages and pages about why one method is better than another. I opted for a simpler approach, explaining the basic concept and then doing t tests with the critical p value being lowered. I wanted not to give any rule of thumb for how much to lower the critical p value, but I knew I would get complaints. So, the rule of thumb is 5% divided by the number of comparisons (not the number of conditions). The reason why I did not want to give a steadfast rule is that the choice of how much to lower the p value depends on how important Type I and Type II errors are. If falsely reporting that an effect exists (a Type I error) would be really bad, then you would want to lower the p value a lot (the 'rule of thumb' value would be about right). However, if failing to detect an effect (a Type II error) is really bad, then you would not want to lower the critical p value by much, and you might even want to increase it.[7]

Particularly in the second example, on Thatcher's resignation, I stressed that statistical significance does not imply that the effect is large or that it is important. This is critical. Too often I see people do statistics as automatons, calculating their F and p values, and not truly understanding their data. They focus too

---

7 I do not actually recommend raising the critical p value, even though an argument could certainly be constructed to justify it in some circumstances. The problem is that it would be very much against the status quo, and it would take a while to convince your readers.

much on the *p*. It is important to look at the data, to graph the data and to think about it. Recall Rosenthal's advice from Chapter 5, that people should 'make friends with their data'. I had a lab 'meeting' one evening over drinks and we were discussing this quote. One of my post-grads said doing statistics was like sipping a fine wine. Sure, you can quickly gulp it down, but to enjoy the aroma, to taste all the intricacies of the wine, you should slowly swirl the glass, breathe its fumes and sip the nectar. Perhaps we were getting a little giddy at this stage of the meeting, but the point is that statistics should be done slowly and carefully, and not hastily.

Exercises

7.1 Barnier and McConkey (1998) gave participants 100 pre-paid postcards and said that they should post one of these every day. Seventeen participants were hypnotised when told this, 17 were told to act as if they have been hypnotised (simulators) and 17 were just told to do this. Their means and standard deviations are given below:

	Hypnotised	Simulator	Control
Mean (in % cards returned)	53.35	46.74	15.64
Standard deviation	37.36	41.03	27.73

(a) Construct, and graph, the 95% confidence intervals for these data.

(b) Below is the ANOVA table, with some numbers missing. Rewrite the table, filling in the missing values.

	Sum of Squares	df	Mean Square	F
Between Groups	XXXXX	2	XXXX	XX
Within Groups	65419	XX	XXXX	
Total	79205	58		

(c) Based on the graph and the ANOVA table, what, in a couple of lines, are your conclusions?

7.2 Suppose a researcher was interested in which animals, of cats, dogs, and fish, made good pets. The researcher went to people's houses, and asked if they had cat(s), dog(s) *or* fish (no one with more than one type of pet could take part in the study). If they had one type of pet, they were asked how good a pet it made, on a 1 to 11 scale, with 11 being very good and 1 being very bad. The data are given in Table 7.3.

Table 7.3 **How good a pet are cats, dogs and fish.**

Cat	Dog	Fish
6	2	4
11	4	2
10	10	0
2	6	3
10	8	2
4	5	5
6	6	4
8	4	6
8	9	4
7	11	5

Table 7.4 **Data on alcohol and suggestibility.**

Condition	n	Mean (out of 35)	Standard deviation	Standard error
Control	13	8.2	4.1	1.14
Low	12	7.9	5.4	1.56
Medium	13	3.7	3.1	0.86
High	13	4.3	4.2	1.16

Note: The condition means and standard deviations are taken directly from Table 2 of Santtila et al. (1999). All other numbers here, and in the ANOVA table, are calculated from these and therefore are slightly different from those in the paper owing to rounding error.

(a) Is there evidence that the ratings differ for cat, dog and fish owners?
(b) From what you found in part (a), does this mean that some pets are better than others?

7.3 There is much concern about witnesses being susceptible to misleading questions. Santtila et al. (1999) conducted a study to see how alcohol affected suggestibility. They had four conditions. A control group ($n = 13$) had a placebo, a glass of liquid containing no alcohol (the outside of the glass had been rubbed with alcohol so it smelled like the drinks for the other conditions). The three experimental conditions had alcoholic drinks: a low-dosage group ($n = 12$) with 0.13 ml of 95% alcohol per kg of body weight, a medium group ($n = 13$) with 0.66 ml and a high group ($n = 13$) with 1.32 ml. They administered a Finnish version of Gudjonsson's (1997) suggestibility scale. High scores mean more suggestible. The data are given in Table 7.4, along with some preliminary calculations.

(a) Draw a graph showing intervals for *both* the standard deviation and the 95% confidence intervals.
(b) Rewrite the ANOVA table below, filling in the missing numbers.

	Sum of Squares	df	Mean Square	F	Sig.
Between Groups	212.57	X	X	3.92	.014
Within Groups	849.48	X	X		
Total	1062.05	X			

7.4 In a couple of sentences, when should you do an ANOVA as opposed to a *t* test?

7.5 When doing multiple comparisons, discuss, in about six lines, the issues involved in choosing a critical *p* value for the individual comparisons.

7.6 Imagine that you are the Statue of Liberty and you are unhappy. What is it that worries you about being the Statue of Liberty? Startup and Davey (2001) asked university undergraduates this. A response might be 'people will be walking around in my head'. Participants were then asked why this worried them, and then why this new worry worried them, and so on. This 'catastrophising interview' continued until the student could come up with no more worries. Startup and Davey recorded the number of worry steps and compared three conditions. Before the interview participants listened to Gyorgy Ligeti's *Lux Aeterna*, Chopin's *Waltzes nos. 11 and 12*, or Vivaldi's *The Four Seasons, Spring*, which put them into a negative, a neutral, or a positive mood, respectively. The authors predicted that the negative group would catastrophise the most. Their data are:

Negative (*Lux Aeterna*)	3, 10, 12, 10, 10, 5, 11, 6, 15, 9, 6, 5, 7, 6, 12
Neutral (*Waltzes nos. 11 and 12*)	4, 2, 4, 6, 9, 4, 8, 5, 5, 9, 4, 3, 2, 5, 4
Positive (*The Four Seasons, Spring*)	4, 11, 4, 2, 4, 6, 1, 2, 16, 3, 7, 3, 2, 3, 2

Is there evidence that the different mood inductions create different amounts of catastrophising?

7.7 There is talk about my local football team, Brighton and Hove Albion, moving to a location right by my university, in the town of Falmer. Falmer is a small town and the proposed location is an area of natural beauty. Not surprisingly, there is some concern. Luckin (2001) conducted a survey of people living in Falmer, of people living about a mile away, and of people living in two sections of Brighton. He hypothesised that people living in Falmer would be the most concerned. He developed a scale to measure concern which could range from 1.00 to 5.00. His data are:

Table 7.5 **Data on concern about a proposed stadium (from Luckin, 2001).**

	Sample size	Mean	Std. dev.	95% CI
Falmer	19	4.00	1.04	3.50, 4.50
Nearby	20	2.36	1.26	1.77, 2.95
Loc. A	20	1.94	1.09	1.43, 2.45
Loc. B	20	1.37	0.64	1.07, 1.67
Total	79	2.40	1.41	2.08, 2.71

In Falmer	5.00, 3.27, 4.18, 3.91, 3.40, 4.64, 1.00, 5.00, 4.36, 4.09, 3.82, 4.91, 4.91, 2.64, 5.00, 3.00, 4.36, 3.55, 5.00
Nearby	3.64, 1.27, 1.82, 4.45, 3.45, 1.09, 1.73, 1.55, 4.00, 1.36, 1.00, 3.45, 1.00, 1.70, 2.55, 1.00, 4.18, 3.27, 3.73, 1.00
In Brighton, location A	1.00, 4.55, 1.00, 1.00, 2.82, 2.27, 1.45, 1.73, 4.55, 1.45, 1.00, 1.27, 2.18, 2.36, 1.09, 1.82, 2.64, 2.64, 1.00, 1.00
In Brighton, location B	1.00, 1.00, 1.00, 1.00, 1.00, 1.00, 1.45, 1.00, 3.27, 1.91, 2.00, 1.00, 1.00, 1.00, 1.00, 2.64, 1.00, 1.55, 1.00, 1.55.

Table 7.5 contains some descriptive statistics that you may find useful.

(a) Graph the means with their 95% confidence intervals.

(b) Conduct an ANOVA on these data to determine if the means differ significantly. Interpret the results.

Further Reading

Field, A. P. (2000). *Discovering Statistics using SPSS for Windows*. London: Sage.
 While this book focuses on conducting statistics using SPSS, it has several excellent chapters on ANOVA and extensions to the basic ANOVA models.

Kirk, R. E. (1999). *Statistics: An introduction*. London: Harcourt Brace.
 If I was going to recommend a 750-page introductory textbook, it would probably be this one. Chapter 14 is a good introduction to ANOVA.

A common situation in all of the sciences is when there are two variables and each of these varies along some kind of scale. The question is, usually, how are these two variables related? The procedures described in this chapter help you to answer this question. The most useful of the procedures is one which you have already been introduced to: the scatterplot. The scatterplot is the basis for the three other procedures: drawing a line, using a couple of equations to choose a line, and the correlation. The aims of this chapter are for you to learn how to interpret data with a scatterplot and to decide whether a certain model – that the relationship can be accounted for by a single straight line – is an acceptable description of the data.

Calculating the Regression Line

Suppose that you are a police officer investigating a car crash that a single witness had seen and you want to know how fast the car was going. All you have is the person's estimate of how fast a car was going and the literature (e.g. Loftus & Palmer, 1974) suggests that a person's estimate of a car's speed can be a poor indicator of the real speed. However, suppose that this is the only information that you had to estimate the actual speed. In his BSc thesis, Peter Wright (no relation) was interested in this problem. Suppose that he asked this eyewitness to estimate the speed of 10 cars because he wanted to see how good the witness was at

Table 8.1 **Data and some preliminary calculations for estimating the speed of cars. I have written Σ for the sum of all the values in that column.**

Estimate (x)	Actual (y)	$x_i - \bar{x}$	$y_i - \bar{y}$	$(x_i - \bar{x})(y_i - \bar{y})$	$(x_i - \bar{x})^2$	pred$_i$	residual$_i$
8	12	− 6.4	− 6.8	43.52	40.96	13.54	− 1.54
8	29	− 6.4	10.2	− 65.28	40.96	13.54	− 4.54
35	24	20.6	5.2	107.12	424.36	27.29	− 3.29
13	16	− 1.4	− 2.8	3.92	1.96	16.09	− 0.09
5	7	− 9.4	− 11.8	110.92	88.36	12.02	− 5.02
7	5	− 7.4	− 13.8	102.12	54.76	13.03	− 8.03
14	24	− 0.4	5.2	− 2.08	0.16	16.60	7.40
22	21	7.6	2.2	16.72	57.76	20.67	0.33
27	26	12.6	7.2	90.72	158.76	23.21	2.79
5	24	− 9.4	5.2	− 48.88	88.36	12.02	11.98
				$\Sigma = 358.8$	$\Sigma = 956.4$		

estimating car speeds. Table 8.1 gives the speeds of cars and the estimates given by the witness, plus some calculations that will be used later in this section.

Figure 8.1 shows a scatterplot with all the actual speeds and the estimates. The mean line for actual speed (the mean was 18.8 mph) of the 10 cars is shown. There are also vertical lines from the mean line to each of the points. If the length of each of these lines is squared (i.e. multiplied by itself), and then these values are all added together, you get what is called the *total sum of squares* (these calculations are done on p. 96). This is a measure of how spread out the data are from the mean line. If this sum is divided by $n - 1$, here 9, you get the variance (approximately 69.5). If this sounds similar to the computations for ANOVA (Chapter 7), it should. The two are closely related. ANOVA is a special case of regression, though showing this is beyond the scope of this book (see Cohen, 1968; Wright, 1997b).

The equations for finding the regression line look complicated. They are, so it is worth being careful when solving them. The mean for both variables should be found first. These are 18.8 mph for actual speed (the y axis variable) and 14.4 mph for estimated speed (the x axis variable). The following are the equations for the slope ($\beta 1$) and the intercept ($\beta 0$) of the regression line:

$$\beta 1 = \frac{\sum (x_i - \bar{x})(y_i - \bar{y})}{\sum (x_i - \bar{x})^2}$$

$$\beta 0 = \bar{y} - \beta 0 \bar{x}$$

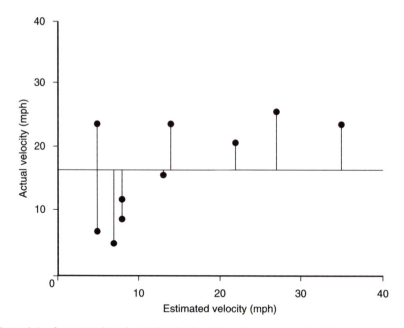

Figure 8.1 **A scatterplot of actual velocity with estimated velocity. The horizontal line shows the mean actual velocity. The vertical lines show the distance from the mean to each data point.**

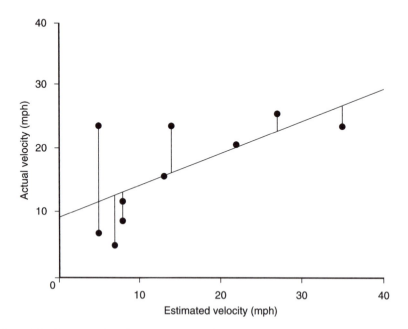

Figure 8.2 **A scatterplot of actual velocity with estimated velocity. The diagonal line shows the regression. The vertical lines show the residuals for each data point.**

Solving for these with the values calculated in Table 8.1 yields $\beta1 = 358.8/956.4 = 0.375$ and $\beta0 = 18.8 - 14.4(0.375) = 24.2$. Thus, the regression equation is

$$\text{Actual}_i = 24.2 + 0.375 \text{ Estimated}_i + e_i$$

Figure 8.2 shows this regression line. As can be seen, there is a positive slope, meaning that as the estimated speed goes up, the actual speed tends also to be larger. This makes sense. The intercept is 24.2 mph, meaning that if this witness thinks the speed of a car is 0 mph, the best estimate using this model is 24.2 mph. This doesn't make sense. With only 10 cases and a large amount of error, the regression line is unreliable, so any interpretation needs to be done carefully.

As with Figure 8.1, in Figure 8.2 I have drawn vertical lines from the model to each point. These residuals are shown in the final column of Table 8.1. It is the difference between the length of these lines and those in Figure 8.1 that shows how much the information about the estimated velocity helps in the prediction.

Let's consider another example. Many questionnaires attempt to measure people's attitudes on a variety of topics. Suppose someone was interested in the relationship between attitudes towards gun control and views on tougher prison sentencing. The researcher might ask a sample of 100 people to rate their agreement on the following two statements:

> 'People should be allowed to buy guns without any restrictions'
> 'The laws on prison sentencing are far too weak'

using a scale from 1 ('totally disagree') to 11 ('totally agree'). Figure 8.3 shows a scatterplot between the responses to these two variables. I have used what are called 'sunflowers' in this graph. Because there are a large number of people in the sample, there are many coordinates, or points, on the scatterplot that have more than one person at them. If there is just a single person at a point then there is just the circle. If there are two people, like the pair giving the response 2 to each question, then the 'sunflower' has two petals. The point above that (2 to the 'guns' question and 3 to the 'sentencing' question) has four people, and therefore four petals are drawn. 'Sunflowers' is a common option in many statistical computing programs (this one was done using SPSS). The regression line is also printed on this graph. It shows a small increase in agreeing with 'sentencing' is associated with an increase in the 'guns' question. It is left as an exercise to find the exact form of this equation.

The Correlation

Sometimes the data points in a scatterplot are close to the regression line, sometimes they are not. It would be useful to have some statistic that measures how close the points are to the regression line, what is sometimes called the 'fit' of the

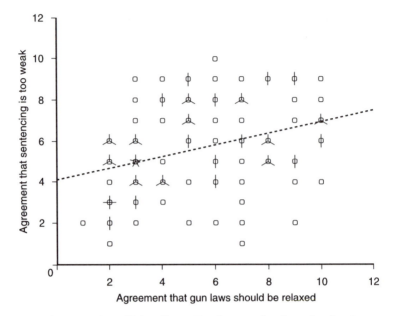

Figure 8.3 **A scatterplot, with 'sunflowers' and regression line, showing the association between views on gun control and sentencing.**

regression line. Not surprisingly, given this section's title, the *correlation* is just such a statistic. It gives a value, ranging from −1 to +1, that measures the fit of a straight line to the data. Negative scores mean that as the values of one variable go up, the values of the other variable tend to go down. Positive scores mean that as the values of one go up, so do the values of the other. Values near either −1 or +1 mean that the regression line has a good fit. Values around 0 mean that the regression line does not fit the data well.

The correlation value is denoted with the letter r. The equation for it is

$$r = \frac{\sum (x_i - \bar{x})(y_i - \bar{y})}{\sqrt{\sum (x_i - \bar{x})^2 \sum (y_i - \bar{y})^2}}$$

If we consider the data in Table 8.1, most of the information needed for the above equation has already been calculated in the table. The only additional measure is $\sum (y_i - \bar{y})^2$. Here are the calculations for that:

$$
\begin{aligned}
\sum (y_i - \bar{y})^2 &= (-6.8)^2 + 10.2^2 + 5.2^2 + (-2.8)^2 + (-11.8)^2 \\
&\quad + (-13.8)^2 + 5.2^2 + 2.2^2 + 7.2^2 + 5.2^2 \\
&= 46.24 + 104.04 + 27.04 + 7.84 + 139.24 \\
&\quad 190.44 + 27.04 + 4.84 + 51.84 + 27.04 \\
&= 625.60
\end{aligned}
$$

These values can be entered into the correlation equation:

$$r = \frac{358.8}{\sqrt{(956.4)(625.6)}} = \frac{358.8}{773.5} = 0.46$$

As this is positive, we know that as the estimated speed goes up, the actual speed also tends to increase. It is often useful to square the correlation: $0.46 \times 0.46 = 0.21$. This value is the amount of variation that the two variables share. It is the proportion of variation in one variance that can be accounted for by variation in the other. If this sounds similar to the R^2 and η^2 found when doing ANOVAs, it should. They are the same.

The next step is usually to see if the observed r differs significantly from 0. In other words, if we assume that in the population the correlation is 0, would we expect to get an r value this large (or larger) less often than 5% of the time (see Chapter 6 for discussion of hypothesis testing). Appendix A gives a table to look up whether the observed value of r exceeds the critical value for $p = 0.01$, $p = 0.05$ and $p = 0.10$ (these are all two-tailed tests). You need to know the sample size to use this table. Here the sample size is 10. You go to the row corresponding to $n = 10$. The critical values are 0.765, 0.639 and 0.549 for $p = 0.01$, $p = 0.05$ and $p = 0.10$, respectively. All of these are larger than the observed value of 0.46. Therefore this correlation is not significantly different from 0. As the sample size increases, the critical values for the correlation become smaller and smaller. If there had been an n of 20, 0.46 would have been significant at $p = 0.05$. As with the other tests discussed in this book, when done on a computer, the computer usually writes a precise p value. You should usually write this value when it is available.

Further Considerations of Regressions and Correlations

There are some additional issues worth considering for regressions and correlations. First, the regressions described here are just for straight lines. In many cases there is a clear relationship, but it is not linear. Figure 8.4 shows some examples where there are clear relationships, but they are not linear. There are more complex regressions that can be used to model these relationships. The most important technique, however, remains using scatterplots to allow you to understand the nature of any relationships.

A second consideration about regressions and correlations is the effect of points that are far away from the others. This includes outliers, those with large

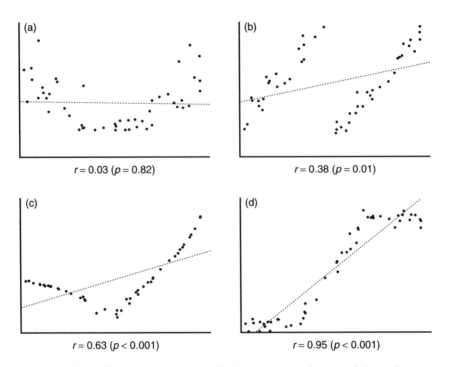

Figure 8.4 **Four different scatterplots with their corresponding correlation values.**

residuals, but also those near the regression line but away from the main group of data. These have a large impact on the regression line and the correlation. It is important to be aware of their impact. I recommend always trying regressions both with and without these points. Figure 8.5 shows a scatterplot with 10 points. The correlation is 0.64. Two of the points are labelled. In the upper left hand corner is an outlier; it does not fit with the pattern of the rest of the data. Without this point the correlation is 0.92. There is also a point in the upper right hand corner. These points, sometimes called *leverage* or *influential* points, also have a large impact on the correlation. If this one is excluded the correlation drops to 0.38. In general, if you throw out one or two points and would reach a different conclusion then it is best to be cautious in any conclusions that you draw.

Another technique that makes these points less influential is to rank the variables, from largest to smallest, and then run a correlation. This is known as Spearman's correlation. The value here is 0.69. It is important when interpreting this value that you remember that this is a correlation of the ranked variables. This procedure is described in detail in Chapter 9.

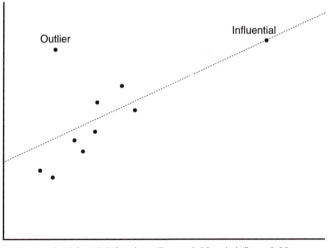

Outlier

Influential

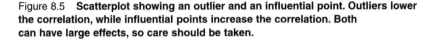

$r = 0.64$ ($p = 0.05$); w/o outlier $r = 0.92$, w/o infl. $r = 0.38$

Figure 8.5 **Scatterplot showing an outlier and an influential point. Outliers lower the correlation, while influential points increase the correlation. Both can have large effects, so care should be taken.**

Finally, there is an often cited adage, 'correlation does not imply causation', which is stated in every introductory statistics book. What often happens after people run a correlation is that they state their conclusions in 'causal' language. So, a researcher might say that increasing the percentage of girls receiving high marks in a local education authority increases the marks for boys. The scatterplot, regressions and correlations described in this chapter do not show this. They simply show that there is a tendency that when an authority has girls who score high, it tends also to have boys who score high. It is often tricky to make sure that you avoid accidentally using causal language inappropriately.[1] Randomly allocating people to conditions is the important precursor to causal conclusions.

Summary

With respect to technical details, this chapter introduced you to the equations for regression and the 'fit' statistic for it, the correlation. Regression is used to plot a

---

1 The phrase 'correlation does not imply causation' is often repeated and stressed in textbook chapters on correlation/regression. But it applies to other procedures when non-experimental methods are used. For example, if comparing the income of Tory and Labour supporters (or Republicans and Democrats in the United States), a *t* test might be used. It is likely that the *t* test would show that the Tories (and Republicans) have higher mean incomes. As far as I know joining these more right wing parties does not cause you to have more wealth. This came up in Exercise 7.2 on pet ownership.

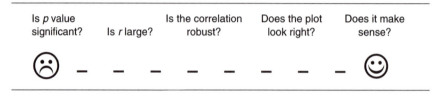

Figure 8.6 **Five ways for assessing the fit of a regression model. All should be used. Using just the ones on the left is discouraged.**

straight line that describes the points in a scatterplot. This can be used to describe the relationship between two variables and for predicting the scores of one variable from another. The correlation is simply a measure of how good the fit of the regression line is. Any time a correlation is reported you should have produced a regression line and a scatterplot. As a correlation is only meaningful with respect to a regression line, and the fit of a regression line is best understood by looking at it with a scatterplot of the data, it is best to start with a scatterplot, run a regression and then look at the correlation.

In assessing the fit of a regression line sometimes people simply see if the correlation is significant, and then say that the regression has a good fit. Figure 8.6 shows that this on its own is not good; it gets the sad face. The next step is to look at the size of the correlation. A correlation can be significant although extremely small in size, simply because the sample is large. Looking just at the correlation and its significance is not enough. As mentioned above, your conclusions should not change too much if a couple of people's data are removed. If they do change, then your regression is not *robust* and you should be cautious in your interpretation (see Figure 8.4). These first three ways to assess the fit of a regression line are all mechanical applications of statistics and are not enough, on their own, to assess the fit of a model.

It is critical to see from the scatterplot if it looks as if the regression line fits the data well. This is subjective, but important. If you were assessing the fit of Figure 8.4(d), it has a large, significant and robust correlation. However, the straight line clearly does not capture the full essence of the data. The final criterion is: does the regression make sense? If, for example, your regression contradicts the law of gravity, the odds are your model is wrong and you should be very cautious with any conclusions. All five of these means of assessment should be used in evaluating the fit of a model, and for all the statistics you use.

Finally, there were a few places where I mentioned that there are more advanced regression techniques. This is an understatement. Regression procedures, of one form or another, underlie almost all statistics. Understanding the basics of regression therefore is a necessary building block for further advancement in statistics.

8.1 Peter Wright actually found that most of his participants were more accurate than the data depicted in Figures 8.1 and 8.2. Figure 8.7 shows the data from one of his actual participants along with the regression line. The following may be useful information:

mean of actual velocity	25.17 mph
mean of estimate velocity	23.05 mph

$$\sum(\text{actual}_i - \overline{\text{actual}})^2 = 3225.37$$

$$\sum(\text{estimate}_i - \overline{\text{estimate}})^2 = 2813.90$$

$$\sum(\text{actual}_i - \overline{\text{actual}})(\text{estimate}_i - \overline{\text{estimate}}) = 2833.37$$

where the 'over-lined' variable names are their means. Find the regression equation for these data. Suppose you were a police officer and this participant had been a witness to a car crash. The driver said he was only driving at 10 mph, but this witness said the car appeared to be going 25 mph. What would you predict is the actual speed from the regression? Do you think the driver's estimate is accurate?

8.2 Zelinsky and Murphy (2000) were interested in the relationship between how long it takes to say the name of an object and the amount of time that a person looks at the object when encoding the object for a recognition test. Suppose participants were shown 10 different pictures of common objects that varied in the amount of time that it takes to name the object, and the researchers recorded the mean time that the participants gazed at the objects. The times, in seconds, that might be found if such a study was run were as given in Table 8.2.

Table 8.2 **Data based on Zelinsky and Murphy (2000).**

Name	Gaze
0.34	0.52
0.37	0.38
0.45	0.50
0.54	0.37
0.59	0.62
0.63	0.53
0.54	0.78
0.83	0.64
0.70	0.93
0.76	0.78

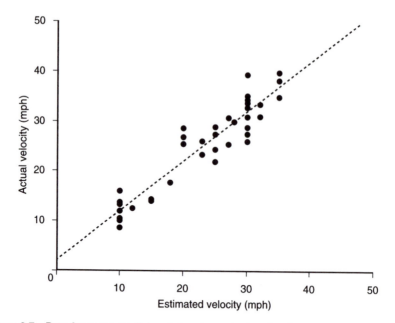

Figure 8.7 **Data from one participant showing actual and estimated vehicle velocities.**

Find whether there is a relationship between these two variables. Draw a scatterplot for these data and add the regression line.

8.3   Find the regression equation shown in Figure 8.3. Let the variable guns$_i$ be denoted x_i and the variable sentencing$_i$ be denoted y_i. The following may be useful information:

$$\text{mean for } x_i = 5.42$$
$$\text{mean for } y_i = 5.65$$
$$\sum(x_i - \overline{x})(y_i - \overline{y}) = 187.70$$
$$\sum(x_i - \overline{x})^2 = 660.36$$
$$\sum(y_i - \overline{y})^2 = 484.75$$
$$n = 100$$

Write out the regression equation and remember to include subscripts where appropriate and a term for the residuals.

8.4   Find the correlation for the 'sentencing' and 'guns' in Exercise 8.3. Is this statistically significant? Is the correlation larger than the correlation for the data in Table 8.1? Are your answers to these questions contradictory?

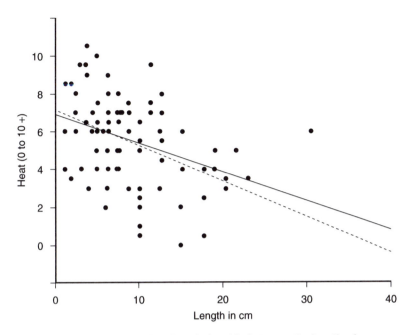

Figure 8.8 **A scatterplot showing the relationship between the length of a chile and the heat. The solid line going through it is the regression line for all the data. The dashed line is the regression line without Nu Mex Big Jim.**

8.5 If cooking with chiles, it is important to recognise that some are much hotter than others. One of my favourite chile databases is on http://easy-web.easynet.co.uk/~gcaselton/chile/chile.html. On it, I found 85 chiles where both the length (in centimetres) and heat (on a 0 to 10 scale, but one of the Habaneros got a $10+$ which I recorded as $10.5)^2$ were listed. The resulting scatterplot is shown in Figure 8.8. The solid line going through the data is the regression line, and has the following equation:

$$Heat_i = 6.86 - 0.15 \, Length_i + e_i$$

(a) Which tend to be hotter, small or large chiles?

---

2 The Habaneros chiles are very hot. These should be avoided as they can cause great discomfort and illness. However, if you do get past the heat, they have an apple-like scent and come in a variety of pretty colours.

(b) If you found a chile that was 12 cm long, how hot on average, on the 0 to 10 + scale, would you expect it to be? From the scatterplot, describe (in words) how confident you would be that the chile's heat was approximately what you found from the equation.

(c) There is an outlier in the upper right area of the graph. Many of you will immediately recognise this as Nu Mex Big Jim, the world's largest chile. It was engineered to be big (and in fact some specimens are much larger than the 30.5 cm average in the graph). As it was engineered to be big, it might be worth re-running the regression without this chile. Will the correlation be larger or smaller after this outlier has been removed?

(d) The regression line without Nu Mex Big Jim is shown by the dashed line on Figure 8.8. Its equation is

$$Heat_i = 7.1 - 0.19\ Length_i + e_i$$

If you came across a 30 cm chile, growing naturally, what heat level would you predict? How confident would you be in the accuracy of this estimate and why?

8.6 One of my pet peeves is signs saying that a garage or driveway is 'In constant use'. I look at it, notice that it is currently not in use, and park my car there. When the owner arrives he (always it's a *he*) is usually ranting and raving that he can't park his car and 'didn't I see the sign?' I reply 'yes, which is why I park there'. Why is the owner in the wrong (and annoying)?

Further Reading

Abelson, R. P. (1995). *Statistics as Principled Argument.* Mahwah, NJ: Lawrence Erlbaum.
    This book, like Figure 8.6, describes how it is important to look at more than just *p* values to convince people that your data have something to say.

Miles, J., & Shevlin, M. (2001). *Applying Regression & Correlation: A guide for students and researchers.* London: Sage.
    This is a really nice book covering the two variable regressions discussed here and more complicated ones.

9 Robust Alternatives for the *t* Tests and the Correlation

Figure 8.6 stressed that your statistical tests should be robust, that omitting a few points should not dramatically alter your conclusions. The statistics that you have learned so far can be affected by a few data points. There are alternatives for both *t* tests and the correlation that are more robust. In fact, several new robust methods are described in statistics journals every year, far too many to describe here. Here you are introduced to three of the most popular robust techniques: the Wilcoxon signed ranks test, the Mann–Whitney *U* test and Spearman's correlation. These are the alternatives for the paired *t* test, the group *t* test and Pearson's correlation, respectively. Each of these involve ranking the data. Briefly, if you have the following 10 scores:

4   7   3   9   14   5   8   10   22   2

the rank of these is

3   5   2   7   9   4   6   8   10   1

Because 2 is the lowest value, it gets rank 1; 3 is the second lowest so gets rank 2, and so on until 22, which is the highest, and gets rank 10. If the data had been

| 3 | 7 | 3 | 9 | 14 | 5 | 8 | 10 | 22 | 2 |

these would get the ranks

| 2.5 | 5 | 2.5 | 7 | 9 | 4 | 6 | 8 | 10 | 1 |

Because two scores have the value 3, the ranking is slightly more complicated. The lowest value is still 2 so it has rank 1. The next lowest is 3. There are two scores with this value. We could arbitrarily give one rank 2 and one rank 3. Instead of doing this we give them the mean of the ranks that we would have arbitrarily assigned to the pair. The mean of 2 and 3 is 2.5. This is called assigning the *mid-rank* of the values. If there were 3 scores with the value 3, the mid-rank of 2, 3 and 4 would be 3.

The *t* tests and correlation made assumptions about the distribution of the variables. These procedures do not make these assumptions. They are often called *distribution-free tests*. They are very useful when there are several outliers and when the distributions are skewed. Many textbooks state that they do not assume that the variables have interval level of measurement (see Chapter 1). This is true for the Mann–Whitney *U* and Spearman's correlation, but not for the Wilcoxon test, which assumes interval data. As I said in Chapter 1, there is much discussion about levels of measurement and how they should affect your choice of statistical test. Providing that it is meaningful to think of a variable as interval, then the choice of which test to use should be based more on its distribution.

The Wilcoxon Test: an Alternative to the Paired *t* Test

One of the assumptions of the paired *t* test was that the difference scores were normally distributed. When they are not normally distributed there are several alternatives. One of the most popular is the Wilcoxon matched pairs signed ranks test, or what I refer to here simply as the Wilcoxon test.[1] This test does not make any distributional assumptions and is therefore referred to as a distribution-free test. It is less affected by extreme points compared with the *t* test. Statisticians say it is more robust. I will go through a short example.

A total of 25 people were asked to retrieve a memory of a happy occasion and of a sad occasion. The response times, in seconds, are shown in the first two columns of Table 9.1. Response times often are not distributed like the normal distribution because they tend to have a few very large times. Therefore, more robust statistics are often used with response times.

---

1 Frank Wilcoxon has another popular test with his name on it, called the Wilcoxon rank sum test. It is used in the same situation as the Mann–Whitney test.

Table 9.1 **Hypothetical data comparing reaction times, in seconds, for retrieving a happy memory and a sad memory.**

Participant	$HAPPY_i$	SAD_i	$DIFF_i$	$RANK_i$	$T+_i$	$T-_i$
1	3.0	10.4	− 6.5	16	−	16
2	10.0	7.1	2.9	6.5	6.5	−
3	7.1	12.1	− 5.0	14	−	14
4	9.7	6.3	3.4	8.5	8.5	−
5	10.0	8.8	1.2	4	4	−
6	9.9	5.1	4.8	13	13	−
7	6.1	14.9	− 8.8	21	−	21
8	18.9	6.6	12.3	22	22	−
9	5.9	8.8	− 2.9	6.5	−	6.5
10	2.8	11.0	− 8.2	19.5	−	19.5
11	3.8	20.4	− 16.6	23	−	23
12	5.7	11.4	− 5.7	15	−	15
13	8.5	7.7	0.8	3	3	−
14	5.4	9.0	− 3.6	10	−	10
15	6.2	14.4	− 8.2	19.5	−	19.5
16	6.0	6.4	− 0.4	2	−	2
17	6.4	14.5	− 8.1	18	−	18
18	9.2	9.1	0.1	1	1	−
19	2.3	9.5	− 7.2	17	−	17
20	7.0	30.6	− 23.6	24	−	24
21	10.1	14.4	− 4.3	12	−	12
22	6.1	9.8	− 3.7	11	−	11
23	7.4	7.4	0.0	−	−	−
24	8.7	5.3	3.4	8.5	− 8.5	−
25	5.2	8.1	− 2.9	5	−	5
					$\sum T+_i = 66.5$	$\sum T-_i = 233.5$

The first step for conducting a Wilcoxon test is the same as when doing a paired t test (see Chapter 6): subtract the scores of one variable from the other. Here I took $HAPPY_i - SAD_i$ to make $DIFF_i$. Most of these scores are negative, which corresponds to slower retrieval times for sad memories than for happy memories. There is one zero difference. People who have the same scores for the two variables, like participant 23, are excluded from the analysis making the effective sample size $n = 24$. They are excluded because they do not provide any information about which set of reaction times is faster. Next, rank the remaining differences by their magnitude, at first *ignoring* whether they are negative or positive. The smallest difference is 0.1 seconds so this is given rank 1. The next smallest is − 0.4 which is given rank 2. Some of the ranks are tied. For example, participant 2 has a difference of positive 2.9 and participant 9 has a difference of negative 2.9. They are tied at the sixth lowest. As there are two of them they could be ranked 6 and 7 if no ties were allowed. The mid-rank of these is 6.5. The ranks are in the fourth column of Table 9.1.

The next step is to separate the ranks for the people who had positive differences from those who had negative differences. I have denoted the positive ranks as $T+_i$ and the negative ranks as $T-_i$. I have included the subscript i to make clear that people can have different values on this variable. Next, add up all the $T+_i$ and all the $T-_i$ to get 66.5 and 233.5 respectively. Either of these can be used in the next equation, but it is usually easier to use the smaller of these, 66.5, and to denote it as just plain T. It is then placed into the rather frightening equation:[2]

$$z = \frac{T - n(n-1)/4}{\sqrt{n(n+1)(2n+1)/24}}$$

Inserting T (66.5) and n (24, since participant 23 is excluded),

$$z = \frac{66.5 - 24(23)/4}{\sqrt{24(25)(49)/24}} = -2.39$$

If we had used $T = 233.5$ we would have found $z = 2.39$ which as you are about to see yields the same p value.

Appendix B is used to find whether a z value is statistically significant. The values in the z columns are the z values. Positive and negative values are treated the same so to find the p value associated with z of -2.39 go to 2.39 in the z column. Then go the the column labelled p. The p value is $p = 0.016$. Since this is less than 0.05 we can say that z of -2.39 is statistically significant. If a z value is greater than 1.96 then it is statistically significant at the 5% level. For the 1% level a z value of 2.58 or greater is necessary for statistically significance.

Mann–Whitney Test: an Alternative to the Between-subjects *t* Test

Helsen and Starkes (1999) ran several very clever studies comparing expert football (soccer) players with novices on a number of tasks. In one of their studies participants watched 30 re-enactments of plays from European and World Cup games on a life-size screen. These were filmed from the vantage

---

2 I am simplifying this procedure a bit. There is a special adjustment that is made when there are many ties (like participants 2 and 9), but usually it does not make a large difference. Also, for small samples there is a special table that can be used (see Siegel & Castellan, 1988, for more details).

Table 9.2   **Data based on Helsen and Starkes (1999). The ranks are done for the entire sample, from 1 to 30.**

Novices		Experts	
Number correct	Rank	Number correct	Rank
8	1.0	22	5.5
9	2.0	23	9.0
17	3.0	23	9.0
20	4.0	24	14.0
22	5.5	25	17.0
23	9.0	26	19.0
23	9.0	26	19.0
23	9.0	27	23.0
24	14.0	27	23.0
24	14.0	27	23.0
24	14.0	28	27.0
24	14.0	28	27.0
26	19.0	29	29.5
27	23.0	29	29.5
27	23.0		
28	27.0		
Sum of ranks = 190.5		Sum of ranks = 274.5	

point of one of the players, whose part the participant was supposed to play. The participant had a football in front of her/him and at a particular point in the film the participant had to shoot, dribble around the goalkeeper or pass to a team-mate. Data based on their study are shown in Table 9.2. Here there are 16 novices and 14 experts.

Before calculating any statistics it is worth graphing these data with a boxplot (see Chapter 2). This is done in Figure 9.1. Two outliers are clear from this figure. There are two novices who perform very poorly. While these could be very influential for a t test, the Mann–Whitney U test is less influenced by these points.

The first step in conducting a Mann–Whitney U test is to rank the variable. Table 9.2 shows the ranks for these data. The lowest number of correct choices was 8, so this value receives rank 1. The second lowest was 9, so it receives rank 2. The third lowest was 17 rank 3, and 20 receives rank 4. There were several ties. There were two participants who got 22 correct, one novice and one expert. These are dealt with by giving the mean rank if the values had been arbitrarily differentiated (i.e. the mid-rank). The most important thing to remember is to do the ranking for the entire sample.

Next, the ranks within each group are added together. Here the sums are 190.5 and 274.5 for the novice and experts, respectively. For notational ease, I will refer to novices as group 1 and experts as group 2, and therefore $n1$ will be the number of novices (16), and $n2$ the number of experts (14), $T1$ is the total sum of ranks

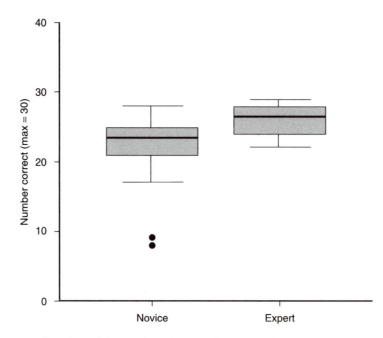

Figure 9.1 **Boxplots of the number of correct football decisions for novices and experts. As there are 30 trials and three possible choices, the two outliers depicted by dots in the novice group are near chance levels. These data are based on Helsen Starkes (1999).**

for novices (190.5) and $T2$ the total sum of ranks for experts (274.5). The Mann–Whitney U is the smaller of

$$\left(n1n2 + \frac{n1(n1 + 1)}{2} - T1\right) \text{ and } \left(n1n2 + \frac{n2(n2 + 1)}{2} - T2\right)$$

For these data these values are

$$(16)(14) + \frac{16(16 + 1)}{2} - 190.5 = 169.5$$

$$\text{and } (16)(14) + \frac{14(14 + 1)}{2} - 274.5 = 54.5$$

Because 54.5 is smaller, $U = 54.5$. As with the Wilcoxon test this value can be changed into a z value. Here the equation is

$$z = \frac{n1n2/2 - U}{\sqrt{(n1n2/12)(n1 + n2 + 1)}}$$

which results in

$$z = \frac{(16)(14)/2 - 54.5}{\sqrt{((16)(14)/12)(16 + 14 + 1)}} = \frac{57.5}{24.06} = 2.39$$

This can be looked up in the z table and as it is the same magnitude as in the last example, it can be rejected at the 5% level.

Spearman's Correlation: an Alternative to Pearson's Correlation

It is not uncommon for people on parole to be economical with the truth when talking with their parole officer. Porter et al. (2000) assessed whether a workshop on detecting deceit could help Canadian parole officers to judge whether someone is telling the truth or not. They found that the parole officers were more accurate after training than before training. One question that can be asked is if the people who did well before the training also did well afterwards. Suppose that, before any training, 20 officers were shown 10 videos of people, either telling the truth or telling lies, and asked whether they thought the person was being honest. Suppose another set of 10 videos were shown after training, and again officers were asked if they thought the person was being honest. Figure 9.2 shows a scatterplot for data that might be collected in these circumstances. They appear to show a positive association. There is one data point that is far away from the others. Someone got 9 out of 10 correct in the initial training and 10 out of 10 correct after training. Pearson's correlation (see Chapter 8) could be run on these data. It gives $r = 0.50$ which has $p = 0.03$. However, if it is run without this point the correlation drops to $r = 0.18$ and the p value becomes $p = 0.46$. Pearson's correlation is very sensitive to points that are far away from the main grouping of the data.

Spearman's correlation, sometimes called Spearman's ρ (the Greek letter rho) or r_s, is less sensitive to individual points. To calculate this correlation you must first rank each of the variables *separately*. When ranking both the pre-training and the post-training variables there are a number of tied ranks, and as with the other procedures you give them the mid-rank. Then you simply calculate Pearson's correlation on the two ranked variables, using the procedures described in Chapter 8. This yields $r_s = 0.31$. This is smaller than the value for Pearson's r, but often it is larger. From Appendix A, the critical value for $n = 20$ and the 5% significance level is 0.44. As the observed value is lower than this, the value is non-significant. If running this procedure, with these data, on a computer the more precise p value is 0.19.

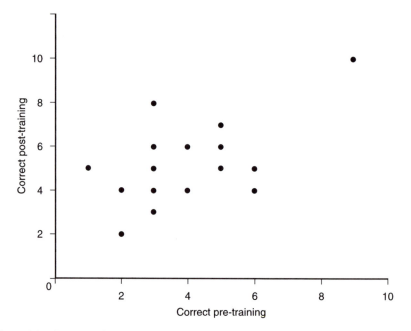

Figure 9.2   **A scatterplot showing the number of videos correctly identified as being truthful or deceitful before training with the number of videos correctly identified after training.**

It is worth noting that when there are a lot of ties then the value you get for r_s is often too large. You should be aware of this and be cautious. There are formulae that correct for this (see Marascuilo & McSweeney, 1977), but these are beyond the scope of this book.

Summary

There is a divide among methods people in psychology. Some say that whenever the assumptions of a test are not met, then tests like those described in this chapter should be used. Others say that providing the distributions do not contain huge outliers or have very non-normal patterns, then the statistics from Chapters 5 to 8 provide a close enough fit. Despite this divide, both groups agree that researchers should be aware of the alternatives and always look at the distributions of the variables.

There is a mistake in a lot of textbooks. Introductory textbooks often say that the distribution-free alternatives are not as powerful as those statistical tests discussed earlier in this book. This would mean that you are less likely to get a significant effect even when an effect is present. While this is true *if* all the assumptions are true, distribution-free alternatives are often *more* powerful when the assumptions are not met. Micceri (1989) looked at a large number of psychology data sets and likened the prevalence of normal distributions to that of unicorns. Therefore, it is safe to say that most distributions are not normal. Further, newer robust tests (see Wilcox, 1998) are even more powerful in many circumstances. Therefore, you should consider robust methods. The one drawback is that they are less common than the *t*, ANOVA and regression statistics. Therefore they appear less often in journals and less often in statistical packages.

The aim of this chapter was to introduce you to distribution-free alternatives for both *t* tests and the correlation. There are further distribution-free tests (one of the most popular textbooks is Siegel & Castellan, 1988). The tests described by Siegel and Castellan, and those described in this book, are often called the traditional distribution-free tests. There is a new breed of robust estimators, like trimmed estimates and M estimates, which are gaining in popularity. Basically, while the *t* tests, ANOVAs, correlations and regressions are greatly influenced by outliers, these newer techniques are less influenced by outliers. The mathematics for these is more complex which makes them less common in the literature, though this may change.

Exercises

9.1 Give the ranks for the following set of data:

$$-4 \quad 12 \quad 4 \quad 0 \quad 2 \quad 2 \quad 2 \quad 8 \quad -1 \quad 1$$

9.2 Using the data in Table 5.1, test the hypothesis that instant and fresh coffee are equally well liked, using the Wilcoxon test. In Exercise 6.4 you used the *t* test to evaluate this hypothesis. What should you consider when deciding which of these two tests is more appropriate?

9.3 Brown and Cline (2001) examined all motor vehicle accidents in part of North Carolina over a couple of years to assess the importance of wearing a seatbelt, and found that wearing a seatbelt greatly reduced the severity of injuries. Suppose injury severity is measured on a 0 to 10 scale where 0 is no injury and 10 is fatal. The following could be data from 20 vehicle occupants:

Wearing seatbelt	4	0	0	1	2	0	7	1	1	2	3	9
Not wearing seatbelt	0	8	9	1	10	0	6	10				

What can you conclude from these data regarding seatbelt use?

Table 9.3 **The relationship between temperature in degrees Celsius and aggression.**

Temperature	Aggressive rating
1	0
−2	6
3	0
11	1
7	0
34	2
4	0
0	0
26	7
14	4
7	1
8	0
18	2
16	2
23	5
1	5
21	6
17	3
21	3
4	0
3	0
11	1
7	0
34	2
4	0
0	0
26	7
14	4
7	1
8	0
18	2
16	2
23	5
1	5
21	6
17	3
21	3
4	0

9.4 There is a growing body of evidence that temperature and violence are associated (see Anderson, 2001, for a review). Suppose that for 10 days a developmental psychologist recorded the temperature in degrees Celsius and rated, from 0 to 10, how aggressive the children were at a primary school playtime. The data are shown in Table 9.3 and in the scatterplot in Figure 9.3. Test whether there is a relationship between temperature and aggression in this setting.

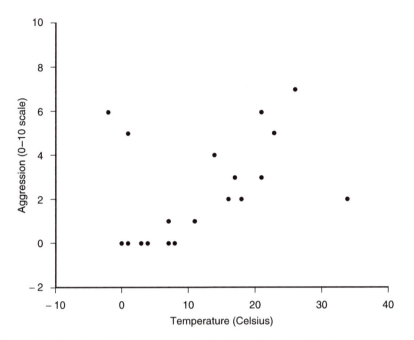

Figure 9.3 **A scatterplot between a psychologist's rating of children's aggressiveness and temperature on a particular day.**

9.5 Students from the University of California at Santa Cruz and at Berkeley were questioned about experiencing the Loma Prieta earthquake and hearing the news that the Bay Bridge had collapsed (Neisser et al., 1996). A year and a half later they were questioned about these events to see how consistent these reports were with their initial reports. Neisser and colleagues created an accuracy scale that could range from 0 to 100. Most of the students had very consistent memories. The scores were negatively skewed. The researchers found that memories were significantly more accurate for experiencing the earthquake than for hearing about the bridge. They also found that the memories were better for the Santa Cruz students than for the Berkeley students. They used two of the statistical tests described in this chapter. Which ones did they use (and for which comparison) and why?

Further Reading

Marascuilo, L. A., & McSweeney, M. (1977). *Nonparametric and distribution-free methods for the social sciences.* Monterey, CA: Brooks/Cole.

This is a book, like Siegel & Castellan, that goes through a large number of distribution-free tests. It goes into a bit more theory about each of the tests and covers more tests.

Siegel, S., & Castellan, N. J. Jr. (1988). *Nonparametric statistics for the behavioral sciences (2nd edn.)* London: McGraw-Hill.
   This is the most used distribution-free textbook. It goes through several tests and shows how to solve them with an example.

Wilcox, R. R. (1998). How many discoveries have been lost by ignoring modern statistical methods? *American Psychologist, 53*, 300–314.
   This is an introduction to modern robust methods, like the trimmed mean. If you want more detail, and more mathematics, I recommend his textbook (Wilcox, 1997).

Contingency Tables, Odds and the χ^2 Test

You have now learned much about the concepts and procedures of statistics. This will have provided you with both the ability to analyse many kinds of data and the foundations to learn more statistics. There is one more set of statistical tests to learn from this book. They are appropriate when you are looking for a relationship between two categorical variables. I will begin with the situation when both variables can take only two possible values. At the end of the chapter a more complex example will be introduced.

Cross-race Identification

A few years ago the football commentator, John Motson, said 'there are teams where you have got players who, from a distance, look almost identical. And, of course, with more black players coming into the game, they would not mind me saying that that can be very confusing' (Radio 5's *Sportsweek*, 4 January 1998). This caused immediate uproar for being politically incorrect, but there is a substantial literature showing that people have relatively more difficulty making cross-race identifications than own-race identifications. However, almost all the research on this topic has used laboratory settings and has shown participants dozens of faces. In most real-world cases people are only asked whether they recognise a single individual. We (Wright et al., 2001) wanted to test the own-race bias hypothesis in a more naturalistic setting.

Table 10.1   **A contingency table, sometimes called crosstabs or cross-tabulation, for the frequency of correct and incorrect choices for each of the four confederates broken down by race of the participant and whether the confederate was identified. The first number is the frequency. The number in parentheses is the percentage.**

Sample	Black confederate		White confederate	
	Blacks	**Whites**	**Blacks**	**Whites**
South Africa				
Number correct	17 (68%)	8 (68%)	15 (60%)	21 (84%)
Number incorrect	8 (32%)	17 (32%)	10 (40%)	4 (16%)
England				
Number correct	19 (95%)	24 (77%)	8 (35%)	14 (52%)
Number incorrect	1 (5%)	7 (23%)	15 (65%)	13 (48%)

In shopping centres in South Africa and England, either a black or a white confederate (someone working for us) went up to either a black or white member of the public. The confederate asked a couple of questions, for example what time it was. A few minutes later the experimenter approached the person and identified herself as a memory researcher. She asked if the person could identify the confederate with whom the participant had just spoken from a set of 10 faces. Table 10.1 gives the number of people in each condition who accurately identified the person. This is called a *contingency table*.

Here these data are analysed separately for each confederate. For the black confederate in South Africa, 17 of the 25 black participants (68%) correctly identified the confederate. It is often useful to think of the *odds* of a correct response. The odds of a correct response are the number of correct responses (17) divided by the number of incorrect responses (8), which is $17/8 = 2.125$. This means that it is about twice as likely that a black participant will make a correct response than an incorrect response. Of the white participants eight correctly identified the confederate while 17 did not. The odds of a correct response for white participants are $8/17 = 0.471$. These data demonstrate the own-race bias: black participants are more accurate than white participants at identifying the black confederate.

A useful statistic to measure the size of this effect is called the *odds ratio* (OR). It is the odds for one group divided by the odds for the other group. Because the hypothesis is that the odds for a correct response should be higher for the black participants (because they are viewing a black confederate) it is better to divide the odds for the black participants by the odds for the white participants, and get $2.125/0.47 = 4.52$, than vice versa. This means that the odds of correctly choosing the confederate are over four times higher if you are a black participant than if you are a white participant. This is a large effect. An OR of 1 would mean that the odds of a correct response were the same for the two groups of participants.

Table 10.2   **Data just for the black confederate in South Africa (Wright et al., 2001), showing the calculations for the odds ratio, denoted OR.**

	Black		White		Calculating the OR		
Correct	A	17	B	8	$OR = \dfrac{A/C}{B/D} = \dfrac{AD}{BD}$		
Incorrect	C	8	D	17			
Odds	$A/C = 2.13$		$B/D = 0.47$		$OR = \dfrac{17/8}{8/17} = \dfrac{17\,(17)}{8\,(8)} = 4.52$		

The data just for this confederate are reprinted with some of the equations in Table 10.2. This is a 2×2 contingency table. In many books this is printed with letters in the different cells, here A, B, C and D. This allows the general equation for the OR to be written. Conceptually it is easiest to think about it as the ratio of two odds, but computationally it is easier to multiply the numbers on one diagonal (A and D) and divide this by the product of the numbers on the other diagonal (B and D). Either method works.

The same calculations can be done for the white confederate in South Africa. The odds of a correct choice for black participants was $15/10 = 1.50$, meaning they were one and a half times more likely to make a correct response than to make an incorrect response. The odds for white participants making a correct response are $21/4 = 5.25$, meaning they were five times more likely to be correct than to be incorrect. Because the confederate was white, the prediction was that the white participants would be more accurate than the black participants. Thus, the OR is $5.25/1.50 = 3.50$. The odds of white participants being correct were 3.50 times higher than for black participants. The odds for these data are shown in Figure 10.1. As can be seen the OR is higher for the black confederate than for the white confederate. Sometimes people just look at the difference between the odds, rather than their ratio. If someone did this, it would make it appear that the effect was larger for the white confederate. It is important to use the ratios. They have several good mathematical properties, although there are some alternatives (see Goodman, 1991; Swets, 1996).

An OR of 1.0 means that the odds of a correct response are the same for both groups. It is often useful to test this hypothesis. The way that this is done is using the χ^2 (chi-squared) test of no association. I mentioned above that A, B, C and D were often used to stand for the four values in Table 10.2. If we let n be the total sample size, the χ^2 value is

$$\chi^2 = \frac{n(AD - BC)^2}{(A+B)(C+D)(A+C)(B+D)}$$

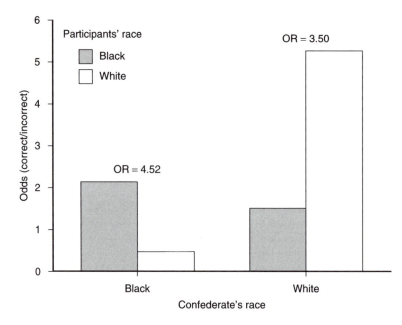

Figure 10.1 **The odds for correctly identifying a black and a white confederate for black and white participants in South Africa (from Table 1 of Wright et al., 2001).**

which yields

$$\chi^2 = \frac{50(289 - 64)^2}{(25)(25)(25)(25)} = \frac{2,531,250}{390,625} = 6.48$$

This value can be looked up in the χ^2 distribution in Appendix E at the back of this book. As with the t and F tests you need to know the degrees of freedom. A handy rule of thumb is to take the number of rows, minus one, and multiply this by the number of columns, minus one. Here, that is $(2 - 1)(2 - 1) = (1)(1) = 1$. For all 2×2 χ^2 tests there is just one degree of freedom. The critical values are 3.84 and 6.63 for the 5% and 1% levels respectively. As the observed value is larger than the critical value at the 5% level we could reject the hypothesis that there was no relationship between a participant's race and accuracy. Examining the data in Table 10.2, and Figure 10.1, the significance is because the black participants were more accurate, which is in line with the prediction. If we were using the more conservative 1% level we would fail to reject this hypothesis.

Table 10.3 **Observed data (O_{ij}) for the white confederate in South Africa (Wright et al., 2001), showing the calculations for expected values (E_{ij}) and the standardised residuals (SR$_{ij}$) for each cell. These can be used to calculate the χ^2 statistic for the entire contingency table.**

	Black	White	Row total (RT$_i$)	Expected values and standardised residuals
Correct	$O_{11} = 15$ $E_{11} = 18$	$O_{12} = 21$ $E_{12} = 18$	RT$_1 = 36$	$E_{ij} = \dfrac{RT_i\,CT_j}{n}$
	SR$_{11} = -0.71$	SR$_{12} = 0.71$		$E_{11} = \dfrac{(36)(25)}{50} = 18$
Incorrect	$O_{21} = 10$	$O_{22} = 4$	RT$_2 = 14$	SR$_{ij} = \dfrac{(O_{ij} - E_{ij})}{\sqrt{E_{ij}}}$
	$E_{21} = 7$ SR$_{21} = 1.13$	$E_{22} = 7$ SR$_{22} = -1.13$		SR$_{11} = \dfrac{-3}{\sqrt{18}} = -0.71$
Column total (CT$_j$)	CT$_1 = 25$	CT$_2 = 25$	$n = 50$	

An alternative method of calculating the χ^2 value, which extends to more complex problems and produces other useful information, is to assume that there is no relationship between the two variables and to calculate the expected values for each cell. We will use this method to calculate the χ^2 value for the white confederate in South Africa. Before showing how this is done it is worth introducing a convenient notation to use with contingency tables. Each cell can be denoted with two numbers placed as subscripts. The first number corresponds to the row number, the second to the column number. For example, the first cell in Table 10.3 shows $O_{11} = 15$. This means that the observed value for the cell in the first row in the first column is 15. The second cell in the first row has the subscript 12 because it is in the first row and the second column. As we have done with other variables, the observed values can be denoted with a general term, O_{ij}, where the subscript i denotes the different rows and the subscript j the different columns.

The first step is to calculate the expected values, denoted E_{ij}, if there was no relationship between the two variables. Overall, 36 of the 50 people, or 72%, correctly chose the confederate. If there were no association between being correct and the participant's race, then you would expect about 72% of black participants and about 72% of white participants to choose the confederate correctly. There are 25 people in each group and 72% of 25 is 18. Therefore, the expected values for both of these groups, for the number correct, is 18. For each, if 18 of 25 are expected to be correct, then seven are expected to be incorrect. An alternative and computationally easier method for calculating the expected value for each cell is to multiply the total number of people in the row by the total number of people in the column and divide by the sample size. This is done in the right hand portion of Table 10.3.

The difference between the observed value and the expected value is the residual. As with the regression procedure, the residual shows how far away the observed value is from the value predicted by the model (here the model is that there is no association between the two variables). It is often useful to show the *standardised residuals*, abbreviated SR. This gives a measure for how far off the expected value is from the observed value, taking into account the row and column totals. It is the residual divided by the square root of the expected value. This is shown in the right hand portion of Table 10.3. The larger the standardised residual for a given cell, the more that cell is responsible for any effect.

Now you are ready to calculate the χ^2 value. It is the sum of the squared standardised residuals:

$$\chi^2 = \sum SR_{ij}^2 = \sum \left(\frac{(O_{ij} - E_{ij})}{\sqrt{E_{ij}}} \right)^2 = \sum \frac{(O_{ij} - E_{ij})^2}{E_{ij}}$$

If you sum the squares of all the standardised residuals in Table 10.3, remembering that the square of a negative number is a positive number, you get

$$(-0.71)^2 + (0.71)^2 + (1.13)^2 + (-1.13)^2 =$$
$$0.50 + 0.50 + 1.28 + 1.28 = 3.56$$

The far right hand side of the above equation shows a more direct way to calculate the χ^2 value. Take each residual, square it, divide by the expected value and then sum these up. For the first cell this is $(15 - 18)^2/18$, 9/18 or 0.50. Either method of calculating these produces the same number.

The big question is: what conclusions should you make? The χ^2 value is 3.56, which is smaller than the critical value of 3.84 found in Appendix E. Therefore it is non-significant. However, from Figure 10.1 it is clear that in this sample the white participants were more accurate at identifying the white confederate than were black participants. The effect is just not large enough given the sample size to be statistically significant. However, as it is in the correct direction and is of a similar magnitude as for the black confederate, it does support the hypothesis (see Berkowitz, 1992, for further arguments about how even non-significant replications lend support to a theory). The analyses of the English data are left as exercises.

Not Propagating the Species: the Darwin Awards

A fundamental tenet of the theory of evolution is that the fittest people are more likely to reproduce, passing their genes on to future generations. Wendy Northcutt (2000)

Table 10.4   **The breakdown, by gender, of the different classifications in Northcutt's (2000) book,** *The Darwin Awards.* **The first number is the observed frequency. The second number is the expected value. The gender of the recipient was unclear in some of the descriptions. Some of these I did not use, while for others I tried to infer the gender.**

	Category				
	Darwin Award	Honorable Mention	Urban Legend	Personal Account	Row total (RT)
Males	131	23	16	35	205
	130.76	23.62	20.25	30.37	
Females	24	5	8	1	38
	24.24	4.38	3.75	5.63	
Column total (CT)	155	28	24	36	243

has explored the converse, that there are people who should, and do, stop their genes being passed on to future generations. People who die (or at least end their chances of reproducing) doing particularly stupid things can be given Darwin Awards. For example, a 1994 winner was tipping a Coke machine over trying to get a free soda when it fell on him and killed him. This meant he could not pass his genes on. Some people do stupid things that do not kill them, like Larry Walters who connected 45 helium-filled weather balloons to his lawn chair, grabbed some Miller Lite beer, and planned to float just above his backyard looking down at his neighbours. Instead, his aircraft shot up. Two pilots from nearby Los Angeles International Airport radioed to report a man drinking beer in their air space. Eventually he came down, was arrested, but still alive and able to reproduce. Therefore he only received Honorable Mention. Northcutt, who began the Darwin website (www.darwinawards.com) while researching neuro-science at Stanford University, also describes Urban Legends, which did not really happen, and Personal Accounts, where people write in about themselves or other people doing stupid things.

Table 10.4 shows the number of males and females in each of these four categories, and the expected values. For example, the expected value for the males with Darwin Awards is

$$E_{11} = \frac{(RT_1)(CT_1)}{n} = \frac{(205)(155)}{243} = 130.76$$

The odds for a Darwin Award winner being male are $131/24 = 5.46$, meaning that a Darwin Award winner is 5.46 times more likely to be male rather than female. It is left as an exercise to calculate the odds of a recipient being male for the other categories.

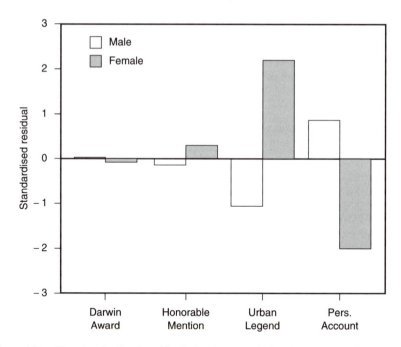

Figure 10.2 **The standardised residuals for the association between gender and classification from *The Darwin Awards*. The figure shows that there are more females in Urban Legends than predicted, and fewer females in the Personal Accounts.**

The standardised residuals can be calculated in the same way as for the 2×2 problem: subtract the expected value from the observed value, then divide this difference by the square root of the expected value. For example, for males being in the Urban Legend category, it is $(16 - 20.25)/\sqrt{20.25} = -0.94$. Figure 10.2 shows the standardised residuals for each cell.

The next step is squaring the standardised residuals and adding them together to find the observed χ^2 of 10.32. The degrees of freedom are the number of rows minus one $(2 - 1 = 1)$, multiplied by the number of columns minus one $(4 - 1 = 3)$, which makes $df = 3$. The critical values at the 5% and 1% levels are 7.81 and 11.34, respectively. The value is significant at the 5% level, though not the 1% level. Using the 5% level, the hypothesis that there is no association between which category someone is in and their gender can be rejected. It is important to realise that this is not saying that males are more likely to be represented in the table. It is clear that they are (see Box 10.1 for the method to show this). These statistics already take into account that males are far overrepresented in the Darwin book.

It is worth noting that the method of squaring the residuals and dividing by the expected value for the cell, and then adding up all of these values, will also produce the correct value. The equation for this is

$$\chi^2 = \sum \frac{(O_{ij} - E_{ij})^2}{E_{ij}}$$

BOX 10.1   THE χ^2 TEST FOR A SINGLE BINARY VARIABLE

Most of the people discussed in Northcutt's book (2000, see p. 254) are male. It is a common situation where a researcher has one binary variable, like gender, and the researcher wants to know if the two values are equally likely. The logic is the same as with the other tests described in this chapter. The first step is to calculate the number of people that the model predicts will be in each cell. If the model was that the two values are equally likely, then the prediction is that half the people should be male and half should be female. As there are 243 people in the sample the prediction is that 121.5 should be male and 121.5 should be female. There will not be exactly 121.5 of either gender, but the prediction if the model is right is that the observed frequencies should be close.

Clearly in this case the model does not fit well. The residuals, standardised residuals, and standardised residuals squared are as given in Table 10.5.

Table 10.5   **Males and females in Northcutt (2000).**

	Residuals	SR	SR^2
Males	83.5	7.58	57.46
Females	− 83.5	− 7.58	57.46
Total	0	0	114.92 = χ^2

The alternative method of squaring the residuals, dividing by the predicted value, and then summing these, also works:

$$\chi^2 = \frac{(205 - 121.5)^2}{121.5} + \frac{(38 - 121.5)^2}{121.5} = \frac{6972.25}{121.5} + \frac{6972.25}{121.5} = 114.77$$

The difference between these numbers is just due to rounding. There is one degree of freedom for this problem. For contingency tables with only one variable the number of degrees of freedom is the number of categories minus one, here $2 - 1 = 1$. This value greatly exceeds the critical value so that the hypothesis that males and females are equally likely to show up in Northcutt's (2000) book can be rejected.

In general, if you have a contingency table, even complex ones with several variables, the basic procedure of calculating expected values and using these to calculate a χ^2 value can be used. Suppose that you were checking someone for extra-sensory perception (ESP). From a normal deck of playing cards you look at a card, and ask the person the suit of the card. You then thoroughly shuffle the deck, and repeat this test. You do this 100 times and the person is right 35 times and wrong 65 times. The question is if this is higher than chance. At first glance some people might think this is lower than chance because it is lower than 50%. However, with four suits, chance guessing predicts only 25 being correct. Thus the residual for correct responses is $35 - 25 = 10$, and is $65 - 75 = -10$ for incorrect responses. Dividing by the square roots of their respective expected values yields $10/\sqrt{25} = 2.00$ and $-10/\sqrt{75} = 1.15$. These are the standardised residuals. If they are squared (yielding 4.00 and 1.32) and summed, this makes $\chi^2 = 5.32$, which is significant with one degree of freedom at the 5% level. Therefore the null hypothesis of no ESP can be rejected.

Summary

This chapter introduced you to statistical procedures for exploring the association between two categorical variables. The main test that psychologists discuss is the χ^2 test. This tests if the observed values are different enough from the expected values to reject the hypothesis that there is no association. If the χ^2 value is large enough, it means that the observed data are further away from the predicted values than you would expect, assuming that there was no association between the variables. In these cases, you reject the null hypothesis and say that there is an association between the two variables.

I stressed two other concepts in this chapter that are often not covered in textbooks. The first is the odds ratio. It is a measure of the size of the effect for 2×2 contingency tables. As psychologists increasingly recognise the importance of reporting effect sizes (Wilkinson et al., 1999), reporting effect sizes in these situations will soon become the norm. There are several measures of effect size that could have been introduced. I chose this one because it is clear to interpret and easier to understand its derivation than other measures.

The second extra concept I stressed was the standardised residual. As with other statistics it is important to look at the residuals to ascertain where any effect may be present. The standardised residual adjusts the value to take into account the row and column totals. For most purposes this is worth doing. As with the odds ratio, there are alternatives I could have introduced. I used the standardised residual because it is closely related to the impact that each cell has on the overall χ^2 value. If you square the standardised residuals, and sum these, you get χ^2.

Final Summary

I hope that you found this book useful. Statistics textbooks differ in their aims and audiences. The main aim of this textbook was to introduce the most common statistical tests used in psychology in a relatively non-threatening manner. The expected audience is people wanting to learn about statistics but not in the detail presented in the 600-page introductory textbooks. As is true with any book, there will be certain biases that the author will have. I have tried not to hide these. I think that the biggest problem for statistics students is not understanding the conceptual aspects. I have tried to stress the commonalities among the procedures so that you can generalise your conceptual knowledge of one statistics procedure to others. All the statistical procedures in this book have the form

$$\text{Data} = \text{model} + \text{residuals}$$

Further, for every statistical test, the fit of the model is based, among other things (see Figure 8.6), on the size of the residuals. If you start thinking with this simple equation in mind, then generalising among the tests that you have learned, and to other tests, will become easier.

I also stressed that graphing, in particular good graphing, is extremely important. A good graph can communicate data extremely efficiently. Equally important, it can convey the information simply to people without statistical expertise. You should always try presenting your data to non-scientists, to see if you have done it clearly. Sometimes I read manuscripts where the authors try to use the longest words possible, and present their ideas in the most complicated manner. These are signs of poor writing.

Finally, ANOVA, regression and contingency tables (the χ^2 test) all lead to more complex procedures. In this book all the procedures were either univariate (one variable), like calculating the mean of a variable, or bivariate (two variables), like calculating the correlation between two variables. The next step is to look at multivariate procedures, where more than two variables are examined. This includes more complex ANOVAs, multiple regression, and the analysis of contingency tables with loglinear models. The steps that you have taken throughout this book should allow you to take on these statistics.

Table 10.6 **For amateur meteorologists in Los Angeles and London, whether they say it is going to rain by whether it does rain.**

	Los Angeles			London		
	Says rain	**Says dry**	**Total**	**Says rain**	**Says dry**	**Total**
It rains	9	48	57	120	61	181
It is dry	52	256	308	118	66	184
Total	61	304	365	238	127	365

Exercises

10.1 Calculate the odds for white English participants and black English participants identifying the black and white confederates (from Table 10.1). For each confederate, calculate the odds ratio. Do these data support the own-race bias?

10.2 Find the odds of being male for those given an Honorable Mention in the Darwin Awards from Table 10.4.

10.3 The two cities I have lived most of my life in are Los Angeles and London. Suppose that you wanted to compare two people for their ability to predict whether it was going to rain or not, and that one person was in Los Angeles and one in London. Table 10.6 shows the days that each predicted rain by whether it rained or not. First, what is the percentage of time each is accurate in their prediction? Second, what is the odds ratio for each of the amateur meteorologists? Discuss why your answers may seem contradictory.

10.4 The Sixth Conference of the European Society for Cognitive Psychology was held at Elsinore, Denmark, in 1993. Martin and Jones (1995) asked delegates which way the queen was facing on the 20 Kroner coin. The correct answer is to the right. Seventy-two participants were asked this and only 21 correctly said that the queen faced to the right. What are the odds of a correct answer? Assuming chance guessing is 50%, is this significantly different from chance?

10.5 In Martin and Jones's (1995) study (Exercise 10.4), half of the participants were Danish and half were not. Of the Danish participants, 10 of 36 correctly said that the queen was facing to the right. Of the non-Danish participants, 11 of 36 said she was facing to the right. What is the odds ratio for the relationship between nationality and the direction that people thought the queen was facing on the 20 Kroner coin? Is this significantly different from 1?

10.6 Kuhn and colleagues (2000) explored whether there was an association between the race of the victim and the suspect in youth homicides in Milwaukee County during the 1990s. Table 10.7 provides data based on

Table 10.7 **Data based on Kuhn and colleagues' (2000) study of youth homicide in Milwaukee, Wisconsin.**

Victim	Race of suspect			
	White	**Black**	**Other**	**Total**
White	35	9	8	52
Black	9	250	1	260
Total	44	259	9	312

their Figure 1 showing the race of the suspect for the 52 homicides of white youths and the 260 homicides of black youths.

What are the odds of a white victim having a white suspect? What are the odds of a black victim having a white suspect? Is there an association between the race of the victim and the race of the suspect? Looking at just the black and white victims and suspects, what is the odds ratio?

10.7 According to Darwin Awards data, males appear much more likely to kill themselves than females. From the biological fact that females can only reproduce a relatively few number of times, but that males can spread their genes much more, why might the overall gender difference in Darwin Awards be good for the species?

Further Reading

Northcutt, W. (2000). *The Darwin Awards: Evolution in Action.* New York: Dutton.
    The UK title is *The Darwin Awards: 150 Bizarre True Stories of how Dumb Humans met their Maker.* Not sure why there is a difference, but this is a very entertaining book. My brother bought it for me last Christmas. I also recommend the webpage, www.darwinawards.com

Wright, D. B. (1997b). *Understanding Statistics: An Introduction for the Social Sciences.* London: Sage.
    This book goes into more detail about the analysis of contingency tables, without getting too complex.

Appendix A

The r Table

n	Significance level		
	1%	**5%**	**10%**
3	1.00	1.00	0.99
4	0.99	0.95	0.90
5	0.96	0.88	0.81
6	0.92	0.81	0.73
7	0.87	0.75	0.67
8	0.83	0.71	0.62
9	0.80	0.67	0.58
10	0.76	0.63	0.55
11	0.73	0.60	0.52
12	0.71	0.58	0.50
13	0.68	0.55	0.48
14	0.66	0.53	0.46
15	0.64	0.51	0.44
16	0.62	0.50	0.43
17	0.61	0.48	0.41
18	0.59	0.47	0.40
19	0.58	0.46	0.39
20	0.56	0.44	0.38
21	0.55	0.43	0.37
22	0.54	0.42	0.36
23	0.53	0.41	0.35
24	0.52	0.40	0.34
25	0.51	0.40	0.34
30	0.46	0.36	0.31
35	0.43	0.33	0.28
40	0.40	0.31	0.26
45	0.38	0.29	0.25
50	0.36	0.28	0.24
60	0.33	0.25	0.21
70	0.31	0.24	0.20
80	0.29	0.22	0.19
90	0.27	0.21	0.17
100	0.26	0.20	0.17
150	0.21	0.16	0.13
200	0.18	0.14	0.12

(Continued)

		Significance level	
n	**1%**	**5%**	**10%**
300	0.15	0.11	0.10
400	0.13	0.10	0.08
500	0.12	0.09	0.07
1000	0.08	0.06	0.05

Table created in SPSS.

In order to find if your observed *r* value has reached statistical significance at $p = 1\%$, $p = 5\%$ or $p = 10\%$, you first need to know the number of people in your sample. Go to the first column, labelled *n*, and go down to the appropriate number. If your number is not listed, go to the one above it in the table. For example, if you had 47 participants in your study, you would go to the row with $n = 45$. In order to be significant at the 1% level you need $r = 0.38$ or higher. To be significant at the 5% level you need $r = 0.29$ or higher.

Appendix B
The Normal (z) Distribution

z	p	z	p	z	p
0.00	1.000	1.19	0.234	1.58	0.114
0.05	0.960	1.20	0.230	1.59	0.112
0.10	0.920	1.21	0.226	1.60	0.110
0.15	0.881	1.22	0.222	1.61	0.107
0.20	0.841	1.23	0.219	1.62	0.105
0.25	0.803	1.24	0.215	1.63	0.103
0.30	0.764	1.25	0.211	1.64	0.101
0.35	0.726	1.26	0.208	1.65	0.099
0.40	0.689	1.27	0.204	1.66	0.097
0.45	0.653	1.28	0.201	1.67	0.095
0.50	0.617	1.29	0.197	1.68	0.093
0.55	0.582	1.30	0.194	1.69	0.091
0.60	0.549	1.31	0.190	1.70	0.089
0.65	0.516	1.32	0.187	1.71	0.087
0.70	0.484	1.33	0.184	1.72	0.085
0.75	0.453	1.34	0.180	1.73	0.084
0.80	0.424	1.35	0.177	1.74	0.082
0.85	0.395	1.36	0.174	1.75	0.080
0.90	0.368	1.37	0.171	1.76	0.078
0.95	0.342	1.38	0.168	1.77	0.077
1.00	0.317	1.39	0.165	1.78	0.075
1.01	0.312	1.40	0.162	1.79	0.073
1.02	0.308	1.41	0.159	1.80	0.072
1.03	0.303	1.42	0.156	1.81	0.070
1.04	0.298	1.43	0.153	1.82	0.069
1.05	0.294	1.44	0.150	1.83	0.067
1.06	0.289	1.45	0.147	1.84	0.066
1.07	0.285	1.46	0.144	1.85	0.064
1.08	0.280	1.47	0.142	1.86	0.063
1.09	0.276	1.48	0.139	1.87	0.061
1.10	0.271	1.49	0.136	1.88	0.060
1.11	0.267	1.50	0.134	1.89	0.059
1.12	0.263	1.51	0.131	1.90	0.057
1.13	0.258	1.52	0.129	1.91	0.056
1.14	0.254	1.53	0.126	1.92	0.055
1.15	0.250	1.54	0.124	1.93	0.054
1.16	0.246	1.55	0.121	1.94	0.052
1.17	0.242	1.56	0.119	1.95	0.051
1.18	0.238	1.57	0.116	1.96	0.050

(Continued)

z	p	z	p	z	p
1.97	0.049	2.21	0.027	2.45	0.014
1.98	0.048	2.22	0.026	2.46	0.013
1.99	0.047	2.23	0.026	2.47	0.013
2.00	0.046	2.24	0.025	2.48	0.012
2.01	0.044	2.25	0.024	2.49	0.012
2.02	0.043	2.26	0.024	2.50	0.012
2.03	0.042	2.27	0.023	2.51	0.012
2.04	0.041	2.28	0.023	2.52	0.012
2.05	0.040	2.29	0.022	2.53	0.011
2.06	0.039	2.30	0.021	2.54	0.011
2.07	0.038	2.31	0.021	2.55	0.011
2.08	0.038	2.32	0.020	2.56	0.010
2.09	0.037	2.33	0.020	2.57	0.010
2.10	0.036	2.34	0.020	2.58	0.010
2.11	0.035	2.35	0.019	2.59	0.010
2.12	0.034	2.36	0.018	2.60	0.009
2.13	0.033	2.37	0.017	2.70	0.007
2.14	0.032	2.38	0.016	2.80	0.005
2.15	0.032	2.39	0.016	2.90	0.004
2.16	0.031	2.40	0.016	3.00	0.003
2.17	0.030	2.41	0.015	3.50	<0.001
2.18	0.029	2.42	0.015	4.00	<0.001
2.19	0.029	2.43	0.014	4.50	<0.001
2.20	0.028	2.44	0.014	5.00	<0.001

This table was produced by the author using SYSTAT's Data Basic.

Suppose we were adopting a significance level of 0.05 or 5%. This is the conventional level adopted by many researchers, but there is nothing particularly special about it (except that it has become a convention). We would need to find a z either greater than 1.96 or less than -1.96 to reject the hypothesis. Only the positive values are shown in the table. The 5% is based on there being 2.5% of the area under the curve greater than $z = 1.96$ *and* 2.5% less than $z = -1.96$.

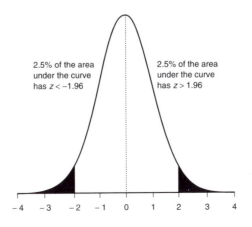

2.5% of the area under the curve has $z < -1.96$

2.5% of the area under the curve has $z > 1.96$

Appendix C
Student's *t*
Distribution

This table was produced by the author using SYSTAT's Data Basic.

df			Critical *p* value			
	0.001	0.010	0.050	0.100	0.200	0.500
2	31.60	9.92	4.30	2.92	1.89	0.82
3	12.92	5.84	3.18	2.35	1.64	0.76
4	8.61	4.60	2.78	2.13	1.53	0.74
5	6.87	4.03	2.57	2.02	1.48	0.73
6	5.96	3.71	2.45	1.94	1.44	0.72
7	5.41	3.50	2.36	1.89	1.41	0.71
8	5.04	3.36	2.31	1.86	1.40	0.71
9	4.78	3.25	2.26	1.83	1.38	0.70
10	4.59	3.17	2.23	1.81	1.37	0.70
11	4.44	3.11	2.20	1.80	1.36	0.70
12	4.32	3.05	2.18	1.78	1.36	0.70
13	4.22	3.01	2.16	1.77	1.35	0.69
14	4.14	2.98	2.14	1.76	1.35	0.69
15	4.07	2.95	2.13	1.75	1.34	0.69
16	4.01	2.92	2.12	1.75	1.34	0.69
17	3.97	2.90	2.11	1.74	1.33	0.69
18	3.92	2.88	2.10	1.73	1.33	0.69
19	3.88	2.86	2.09	1.73	1.33	0.69
20	3.85	2.85	2.09	1.72	1.33	0.69
21	3.82	2.83	2.08	1.72	1.32	0.69
22	3.79	2.82	2.07	1.72	1.32	0.69
23	3.77	2.81	2.07	1.71	1.32	0.69
24	3.75	2.80	2.06	1.71	1.32	0.68
25	3.73	2.79	2.06	1.71	1.32	0.68
30	3.65	2.75	2.04	1.70	1.31	0.68
35	3.59	2.72	2.03	1.69	1.31	0.68
40	3.55	2.70	2.02	1.68	1.30	0.68
45	3.52	2.69	2.01	1.68	1.30	0.68
50	3.50	2.68	2.01	1.68	1.30	0.68
60	3.46	2.66	2.00	1.67	1.30	0.68
70	3.44	2.65	1.99	1.67	1.29	0.68
80	3.42	2.64	1.99	1.66	1.29	0.68
90	3.40	2.63	1.99	1.66	1.29	0.68
100	3.39	2.63	1.98	1.66	1.29	0.68
∞(z)	3.29	2.58	1.96	1.64	1.15	0.67

Suppose you were comparing the means of two groups of people and did not know the population standard deviations. The usual test in this circumstance is the group t test (Chapter 6). Suppose one group had 28 people in it and the other group had 24. The number of degrees of freedom is $20 + 24 - 2$, which is 42. If you were using $p = 0.05$ for rejecting hypotheses then you would need a t value greater than 2.02. When the exact number of degrees of freedom is not listed, as in this case, most researchers recommend using the higher critical t value. This lessens the chances of a Type I error (rejecting a true hypothesis), but it does increase the chances of a Type II error (failing to reject a false hypothesis).

Appendix D
The *F* Distribution

The top value in each row is the necessary *F* value to reject the hypothesis at $p = 0.05$. The second value is for $p = 0.01$.

df	1	2	3	4	5	6	7	8	9	10	20	50	1000
						Numerator degrees of freedom							
1	161.45	199.50	215.71	224.58	230.16	233.99	236.77	238.88	240.54	241.88	248.01	251.77	254.19
	4052.18	4999.50	5403.35	5624.58	5763.65	5858.99	5928.36	5981.07	6022.47	6055.85	6208.73	6302.52	6362.68
2	18.51	19.00	19.16	19.25	19.30	19.33	19.35	19.37	19.38	19.40	19.45	19.48	19.49
	98.50	99.00	99.17	99.25	99.30	99.33	99.36	99.37	99.39	99.40	99.45	99.48	99.50
3	10.13	9.55	9.28	9.12	9.01	8.94	8.89	8.85	8.81	8.79	8.66	8.58	8.53
	34.12	30.82	29.46	28.71	28.24	27.91	27.67	27.49	27.35	27.23	26.69	26.35	26.14
4	7.71	6.94	6.59	6.39	6.26	6.16	6.09	6.04	6.00	5.96	5.80	5.70	5.63
	21.20	18.00	16.69	15.98	15.52	15.21	14.98	14.80	14.66	14.55	14.02	13.69	13.47
5	6.61	5.79	5.41	5.19	5.05	4.95	4.88	4.82	4.77	4.74	4.56	4.44	4.37
	16.26	13.27	12.06	11.39	10.97	10.67	10.46	10.29	10.16	10.05	9.55	9.24	9.03
6	5.99	5.14	4.76	4.53	4.39	4.28	4.21	4.15	4.10	4.06	3.87	3.75	3.67
	13.75	10.92	9.78	9.15	8.75	8.47	8.26	8.10	7.98	7.87	7.40	7.09	6.89
7	5.59	4.74	4.35	4.12	3.97	3.87	3.79	3.73	3.68	3.64	3.44	3.32	3.23
	12.25	9.55	8.45	7.85	7.46	7.19	6.99	6.84	6.72	6.62	6.16	5.86	5.66
8	5.32	4.46	4.07	3.84	3.69	3.58	3.50	3.44	3.39	3.35	3.15	3.02	2.93
	11.26	8.65	7.59	7.01	6.63	6.37	6.18	6.03	5.91	5.81	5.36	5.07	4.87
9	5.12	4.26	3.86	3.63	3.48	3.37	3.29	3.23	3.18	3.14	2.94	2.80	2.71
	10.56	8.02	6.99	6.42	6.06	5.80	5.61	5.47	5.35	5.26	4.81	4.52	4.32
10	4.96	4.10	3.71	3.48	3.33	3.22	3.14	3.07	3.02	2.98	2.77	2.64	2.54
	10.04	7.56	6.55	5.99	5.64	5.39	5.20	5.06	4.94	4.85	4.41	4.12	3.92
11	4.84	3.98	3.59	3.36	3.20	3.09	3.01	2.95	2.90	2.85	2.65	2.51	2.41
	9.65	7.21	6.22	5.67	5.32	5.07	4.89	4.74	4.63	4.54	4.10	3.81	3.61
12	4.75	3.89	3.49	3.26	3.11	3.00	2.91	2.85	2.80	2.75	2.54	2.40	2.30
	9.33	6.93	5.95	5.41	5.06	4.82	4.64	4.50	4.39	4.30	3.86	3.57	3.37
13	4.67	3.81	3.41	3.18	3.03	2.92	2.83	2.77	2.71	2.67	2.46	2.31	2.21
	9.07	6.70	5.74	5.21	4.86	4.62	4.44	4.30	4.19	4.10	3.66	3.38	3.18
14	4.60	3.74	3.34	3.11	2.96	2.85	2.76	2.70	2.65	2.60	2.39	2.24	2.14
	8.86	6.51	5.56	5.04	4.69	4.46	4.28	4.14	4.03	3.94	3.51	3.22	3.02
15	4.54	3.68	3.29	3.06	2.90	2.79	2.71	2.64	2.59	2.54	2.33	2.18	2.07
	8.68	6.36	5.42	4.89	4.56	4.32	4.14	4.00	3.89	3.80	3.37	3.08	2.88

(Continued)

df	\multicolumn{12}{c}{Numerator degrees of freedom}
	1	**2**	**3**	**4**	**5**	**6**	**7**	**8**	**9**	**10**	**20**	**50**	**1000**
16	4.49	3.63	3.24	3.01	2.85	2.74	2.66	2.59	2.54	2.49	2.28	2.12	2.02
	8.53	0.20	5.20	4.77	4.44	4.20	4.03	3.89	3.78	3.69	3.26	2.97	2.76
17	4.45	3.59	3.20	2.96	2.81	2.70	2.61	2.55	2.49	2.45	2.23	2.08	1.97
	8.40	6.11	5.18	4.67	4.34	4.10	3.93	3.79	3.68	3.59	3.16	2.87	2.66
18	4.41	3.55	3.16	2.93	2.77	2.66	2.58	2.51	2.46	2.41	2.19	2.04	1.92
	8.29	6.01	5.09	4.58	4.25	4.01	3.84	3.71	3.60	3.51	3.08	2.78	2.58
19	4.38	3.52	3.13	2.90	2.74	2.63	2.54	2.48	2.42	2.38	2.16	2.00	1.88
	8.18	5.93	5.01	4.50	4.17	3.94	3.77	3.63	3.52	3.43	3.00	2.71	2.50
20	4.35	3.49	3.10	2.87	2.71	2.60	2.51	2.45	2.39	2.35	2.12	1.97	1.85
	8.10	5.85	4.94	4.43	4.10	3.87	3.70	3.56	3.46	3.37	2.94	2.64	2.43
25	4.24	3.39	2.99	2.76	2.60	2.49	2.40	2.34	2.28	2.24	2.01	1.84	1.72
	7.77	5.57	4.68	4.18	3.85	3.63	3.46	3.32	3.22	3.13	2.70	2.40	2.18
30	4.17	3.32	2.92	2.69	2.53	2.42	2.33	2.27	2.21	2.16	1.93	1.76	1.63
	7.56	5.39	4.51	4.02	3.70	3.47	3.30	3.17	3.07	2.98	2.55	2.25	2.02
40	4.08	3.23	2.84	2.61	2.45	2.34	2.25	2.18	2.12	2.08	1.84	1.66	1.52
	7.31	5.18	4.31	3.83	3.51	3.29	3.12	2.99	2.89	2.80	2.37	2.06	1.82
50	4.03	3.18	2.79	2.56	2.40	2.29	2.20	2.13	2.07	2.03	1.78	1.60	1.45
	7.17	5.06	4.20	3.72	3.41	3.19	3.02	2.89	2.78	2.70	2.27	1.95	1.70
75	3.97	3.12	2.73	2.49	2.34	2.22	2.13	2.06	2.01	1.96	1.71	1.52	1.35
	6.99	4.90	4.05	3.58	3.27	3.05	2.89	2.76	2.65	2.57	2.13	1.81	1.53
100	3.94	3.09	2.70	2.46	2.31	2.19	2.10	2.03	1.97	1.93	1.68	1.48	1.30
	6.90	4.82	3.98	3.51	3.21	2.99	2.82	2.69	2.59	2.50	2.07	1.74	1.45
1000	3.85	3.00	2.61	2.38	2.22	2.11	2.02	1.95	1.89	1.84	1.58	1.36	1.11
	6.66	4.63	3.80	3.34	3.04	2.82	2.66	2.53	2.43	2.34	1.90	1.54	1.16

To find whether an F value is significant you need to know both the degrees of freedom of the numerator and the degrees of freedom of the denominator. Usually these are the degrees of freedom of the model and of the residuals, respectively, but for certain hypotheses this need not be the case. Suppose for 45 subjects that we had divided them into five groups. Our model would therefore have four degrees of freedom (four dummy variables to represent the five values). The residuals would have 40 degrees of freedom. The critical values are 2.61 and 3.83 for $p = 0.05$ and $p = 0.01$, respectively.

Appendix E
The χ^2 Distribution

This table was produced by the author using SYSTAT's Data Basic.

	critical level				critical level		
df	0.10	0.05	0.01	*df*	0.10	0.05	0.01
1	2.71	3.84	6.63	23	32.01	35.17	41.64
2	4.61	5.99	9.21	24	33.20	36.42	42.98
3	6.25	7.81	11.34	25	34.38	37.65	44.31
4	7.78	9.49	13.28	30	40.26	43.77	50.89
5	9.24	11.07	15.09	35	46.06	49.80	57.34
6	10.64	12.59	16.81	40	51.81	55.76	63.69
7	12.02	14.07	18.48	45	57.51	61.66	69.96
8	13.36	15.51	20.09	50	63.17	67.50	76.15
9	14.68	16.92	21.67	60	74.40	79.08	88.38
10	15.99	18.31	23.21	70	85.53	90.53	100.43
11	17.28	19.68	24.72	80	96.58	101.88	112.33
12	18.55	21.03	26.22	90	107.51	113.15	124.12
13	19.81	22.36	27.69	100	118.50	124.34	135.81
14	21.06	23.68	29.14	200	226.02	233.99	249.45
15	22.31	25.00	30.58	300	331.79	341.39	359.91
16	23.54	26.30	32.00	400	436.65	447.63	468.73
17	24.77	27.59	33.41	500	540.93	553.13	576.50
18	25.99	28.87	34.81	600	644.80	658.09	683.52
19	27.20	30.14	36.19	700	748.36	762.66	789.98
20	28.41	31.41	37.57	800	851.67	866.91	895.99
21	29.62	32.67	38.93	900	954.78	970.90	1001.63
22	30.81	33.92	40.29	1000	1057.72	1074.68	1106.97

All probabilities in this table refer to the extreme tail (see figure). Suppose you were analysing some categorical data and there were five degrees of freedom in the residuals. This might happen if you were looking to see if there was an association in a 2×6 contingency table (one variable has two possible values, the other six). If you were using the $p = 0.05$ level then you would need a χ^2 value

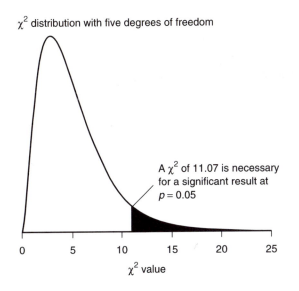

χ^2 distribution with five degrees of freedom

A χ^2 of 11.07 is necessary for a significant result at $p = 0.05$

χ^2 value

of 11.07. The area beyond this point on the graph is 5% of the total area under the curve. Only 5% of the time would you expect a value this high (or higher) if there is no association in the population (providing proper sampling and procedures are used).

References

Abelson, R. P. (1995). *Statistics as Principled Argument*. Mahwah, NJ: Lawrence Erlbaum.

Allen, J. J. B., Schnyer, R. N., & Hitt, S. K. (1998). The efficacy of acupuncture in the treatment of major depression in women. *Psychological Science, 9*, 397–401.

Allen, R. E. (1985). *The Oxford Dictionary of Current English*. Oxford: Oxford University Press.

Anderson, C. A. (2001). Heat and violence. *Current Directions in Psychological Science, 10*, 33–38.

Barnier, A. J., & McConkey, K. M. (1998). Posthypnotic responding away from the hypnotic setting. *Psychological Science, 9*, 256–262.

Berkowitz, L. (1992). Some thoughts about conservative evaluations of replications. *Personality and Social Psychology Bulletin, 18*, 319–324.

Brown, C. K., & Cline, D. M. (2001). Factors affecting injury severity to rear-seated occupants in rural motor vehicle crashes. *American Journal of Emergency Medicine, 19*, 93–98.

Caine, T. M., Foulds, G. A., & Hope, K. (1967). *Manual of the Hostility and Direction of Hostility Questionnaire*. London: London University Press.

Cohen, J. (1968). Multiple regression as a general data-analytic system. *Psychological Bulletin, 70*, 426–443.

Cohen, J. (1990). Things I have learned (so far). *American Psychologist, 45*, 1304–1312.

Cohen, J. (1994). The Earth is round ($p < .05$). *American Psychologist, 49*, 997–1003.

Cook, T. D., & Campbell, D. T. (1979). *Quasi-experimentation: Design & Analysis Issues for Field Settings*. London: Houghton Mifflin.

Cuc, A., & Hirst, W. (2001). Implicit theories and context in personal recollection: Romanians' recall of their political and economic past. *Applied Cognitive Psychology, 15*, 45–60.

Dawes, R. M. (1994). *House of Cards: Psychology and Psychotherapy built on Myth*. New York: Free Press.

Dorling, D., & Simpson, L. (Eds.) (1999). *Statistics in Society: The Arithmetic of Politics*. London: Arnold.

Dyson, F. W., Eddington, A. S., & Davidson, C. (1920). A determination of the deflection of light by the Sun's gravitational field, from observations made at the total eclipse of May 29, 1919. *Philosophical Transactions of the Royal Society A, 220*, 291–334.

Eddington, A. S. (1920). *Space, Time and Gravitation: An Outline of the General Relativity Theory*. Cambridge: Cambridge University Press.

Efron, B., & Gong, G. (1983). A leisurely look at the bootstrap, the jackknife, and cross-validation. *American Statistician, 37*, 36–48.

Everitt, B. S. (1999). *Making Sense of Statistics in Psychology. A Second-Level Course*. Oxford: Oxford University Press.

Festinger, L., & Carlsmith, J. M. (1959). Cognitive consequences of forced compliance, *Journal of Abnormal and Social Psychology*, *58*, 203–210.

Field, A. P. (2000). *Discovering Statistics using SPSS for Windows*. London: Sage.

Fienberg, S. E. (1971). Randomization and social affairs: The 1970 draft lottery. *Science*, *171*, 255–261.

Fisher, R. P., & Geiselman, R. E. (1992). *Memory-enhancing Techniques for Investigative Interviewing: The Cognitive Interview*. Springfield, IL: C. C. Thomas.

Gaskell, G. D., Wright, D. B., & O'Muircheartaigh, C. A. (1993). Measuring scientific interest: The effect of knowledge questions on interest ratings. *Journal for the Public Understanding of Science*, *2*, 39–57.

Goldstein, H. (1995). *Multilevel Statistical Methods* (2nd ed.). London: Edward Arnold.

Goodman, L. A. (1991). Models, measures, and graphical displays in the analysis of contingency tables (with discussion). *Journal of the American Statistical Association*, *86*, 1085–1138.

Gudjonsson, G. H. (1997). *The Gudjonsson Suggestibility Scales Manual*. Hove: Psychology Press.

Harlow, L. L., Mulaik, S. A., & Steiger, J. H. (Eds.) (1997). *What if There were no Significance Tests?* London: Lawrence Erlbaum.

Helsen, W. F., & Starkes, J. L. (1999). A multidimensional approach to skilled perception and performance in sport. *Applied Cognitive Psychology*, *13*, 1–27.

Hennessy, M. B., Davis, H. N., Williams, M. T., Mellott, C., & Douglas, C. W. (1997). Plasma cortisol levels of dogs in a country animal shelter. *Physiology and Behavior*, *62*, 485–490.

Inzlicht, M., & Ben-Zeev, T. (2000). A threatening intellectual environment: Why females are susceptible to experiencing problem-solving deficits in the presence of males. *Psychological Science*, *11*, 365–371.

Keating, J. P., & Scott, D. W. (1999). A primer on density estimation for the great home run race of '98. *Stats*, Spring, 16–22.

Kirk, R. E. (1999). *Statistics: An Introduction*. London: Harcourt Brace.

Kuhn, E. M., Nie, C. L., O'Brien, M. E., Withers, R. L., & Hargarten, S. W. (2000). Victim and perpetrator characteristics for firearm-related homicides of youth during 1991–1997. In P. H. Blackman, V. L. Leggett, B. L. Olson & J. P. Jarvis (Eds.), *The Varieties of Homicide and its Research: Proceedings of the 1999 meeting of the Homicide Research Working Group*. Washington, DC: Federal Bureau of Investigation.

Loftus, E. F., & Palmer, J. C. (1974). Reconstruction of automobile destruction: An example of the interaction between language and memory. *Journal of Verbal Learning and Verbal Behavior*, *13*, 585–589.

Lord, F. M. (1953). On the statistical treatment of football numbers. *American Psychologist*, *8*, 750–751.

Luckin, J. (2001). An exploration into the effect a proposed football stadium has on the local community. Unpublished manuscript.

Maccallum, F., McConkey, K. M., Bryant, R. A., & Barnier, A. J. (2000). Specific autobiographical memory following hypnotically induced mood state. *International Journal of Clinical and Experimental Hypnosis*, *48*, 361–373.

Marascuilo, L. A., & McSweeney, M. (1977). *Nonparametric and Distribution-free Methods for the Social Sciences*. Monterey, CA: Brooks/Cole.

Martin, M., & Jones, G. V. (1995). Danegeld remembered: Taxing further the coin head illusion. *Memory*, *3*, 97–104.

Meehl, P. (1978). Theoretical risks and tabular asterisks. Sir Karl, Sir Ronald and the slow progress of soft psychology. *Journal of Consulting and Clinical Psychology*, *46*, 806–834.

Melhuish, E. C. (1991). Research on day care for young children in the United Kingdom. In E. C. Melhuish and P. Moss (Eds.), *Day Care for Young Children: International Perspectives* (pp. 142–160). London: Routledge.

Micceri, T. (1989). The unicorn, the normal curve, and other improbable creatures. *Psychological Bulletin*, *105*, 156–166.

Miles, J., & Shevlin, M. (2001). *Applying Regression & Correlation: A Guide for Students and Researchers*. London: Sage.

Neisser, U., Winograd, E., Bergman, E. T., Schreiber, C. A., Palmer, S. E., & Weldon, M. S. (1996). Remembering the Earthquake: Direct experience *vs.* hearing the news. *Memory*, *4*, 337–357.

Newton, M. (1998). Changes in measures of personality, hostility and locus of control during residence in a prison therapeutic community. *Legal and Criminological Psychology*, *3*, 209–223.

Northcutt, W. (2000). *The Darwin Awards: Evolution in Action*. New York: Dutton.

Porter, S., Woodworth, M., & Birt, A. R. (2000). Truth, lies, and videotape: An investigation of the ability of federal parole officers to detect deception. *Law and Human Behavior*, *24*, 643–658.

Przibram, K. (Ed.) (1967). *Letters on Wave Mechanics* (trans. M. J. Klein). London: Vision Press.

Reeve, D. K., & Aggleton, J. P. (1998). On the specificity of expert knowledge about a soap opera: An everyday story of farming folk. *Applied Cognitive Psychology*, *12*, 35–42.

Rothblum, E. D., & Factor, R. (2001). Lesbians and their sisters as a control group: Demographic and mental health factors. *Psychological Science*, *12*, 63–69.

Santtila, P., Ekholm, M., & Niemi, P. (1999). The effects of alcohol on interrogative suggestibility: The role of state-anxiety and mood states as mediating factors. *Legal and Criminological Psychology*, *4*, 1–13.

Schkade, D., Sunstein, C. R., & Kahneman, D. (2000). Deliberating about dollars: The severity shift. *Columbia Law Review*, *100*, 1139–1175.

Schreiber, L. R. (1998). *Race for the Record, The Official Major League Baseball Commemorative Book*. New York: Harper Collins.

Schrödinger, E. (trans. J. D. Trimmer) (1983). The present situation in quantum mechanics: A translation of Schrödinger's 'Cat Paradox' paper. In J. A. Wheeler & W. H. Zurek (Eds.), *Quantum Theory and Measurement* (pp. 152–167). Princeton, NJ: Princeton University Press. (Original German publication, Schrödinger, E. (1935). Die gegenwärtige Situation in der Quantenmechanik. *Naturwissenschaften*, *23*, 807–812, 823–828, 844–849.)

Siegel, S., & Castellan, N. J. Jr. (1988). *Nonparametric Statistics for the Behavioral Sciences* (2nd ed.). London: McGraw-Hill.

Startup, H. M., & Davey, G. C. L. (2001). Mood-as-input and catastrophic worrying. *Journal of Abnormal Psychology*, *110*, 83–96.

Stevens, J. (1996). Applied multivariate statistics for the social sciences (3rd ed.). Mahwah, NJ: Lawrence Erlbaum.

'Student' (1931). The Lanarkshire milk experiment. *Biometrika*, *23*, 398–406.

Swets, J. A. (1996). *Signal Detection Theory and ROC Analysis in Psychology and Diagnostics: Collected Papers*. Mahwah, NJ: Lawrence Erlbaum.

Tatar, M. (1998). Teachers as significant others: Gender differences in secondary school pupils' perceptions. *British Journal of Educational Psychology, 68*, 217–227.

Toothaker, L. E. (1993). *Multiple Comparison Procedures*. Newbury Park, CA: Sage.

Wainer, H. (1984). How to display data badly. *American Statistician, 38*, 137–147.

Wainer, H., & Velleman, P. F. (2001). Statistical graphics: Mapping the pathways of science. *Annual Review of Psychology, 52*, 305–335.

Wallgren, A., Wallgren, B., Persson, R., Jorner, U., & Haagland, J. A. (1996). *Graphing Statistics & Data: Creating Better Charts*. London: Sage.

Wilcox, R. R. (1997). *Introduction to Robust Estimation and Hypothesis Testing*. San Diego, CA: Academic Press.

Wilcox, R. R. (1998). How many discoveries have been lost by ignoring modern statistical methods? *American Psychologist, 53*, 300–314.

Wilkinson, L. (2000). Cognitive science and graphic design. In *SYSTAT® 10 Graphics* (pp. 1–18). Chicago: SPSS Inc.

Wilkinson, L. and the Task Force on Statistical Inference, APA Board of Scientific Affairs (1999). Statistical methods in psychology journals: Guidelines and explanations. *American Psychologist, 54*, 594–604.

Wright, D. B. (1997a). Football standings and measurement levels. *The Statistician: Journal of the Royal Statistical Society Series D, 46*, 105–110.

Wright, D. B. (1997b). *Understanding Statistics: An Introduction for the Social Sciences*. London: Sage.

Wright, D. B. (1998a). Modelling clustered data in autobiographical memory research: The multilevel approach. *Applied Cognitive Psychology, 12*, 339–357.

Wright, D. B. (1998b). People, materials and situations. In J. A. Nunn (Ed.), *Laboratory Psychology* (pp. 97–116). Hove: Lawrence Erlbaum.

Wright, D. B. (1999). Science, statistics and three 'Psychologies'. In D. Dorling & L. Simpson (Eds.), *Statistics in Society: The Arithmetic of Politics* (pp. 62–70). London: Arnold.

Wright, D. B., & Loftus, E. F. (1999). Measuring dissociation: Comparison of alternative forms of the Dissociative Experiences Scale. *American Journal of Psychology, 112*, 497–519.

Wright, D. B., Gaskell, G. D., & O'Muircheartaigh, C. A. (1998). Flashbulb memory assumptions: Using national surveys to explore cognitive phenomena. *British Journal of Psychology, 89*, 103–122.

Wright, D. B., Boyd, C. E., & Tredoux, C. G. (2001). A field study of own-race bias in South Africa and England. *Psychology, Public Policy, and Law, 7*, 119–133.

Zelinsky, G. J., & Murphy, G. L. (2000). Synchronizing visual and language processing: An effect of object name length on eye movements. *Psychological Science, 11*, 125–131.

Index